COMPLETE
UNDERCAR
SYSTEMS

Student
Technician's
Shop
Manual

 Delmar Publishers Inc.®

NOTICE TO THE READER

Text, Design, and Production by
Scharff Associates, Ltd.
RD 1 Box 276
New Philadelphia Road
New Ringgold, PA 17960

Scharff Staff
Production Manager: Marilyn Strouse-Hauptly
Cover Design: Eric Schreader
Logo Design: Ed Foulk

Delmar Staff
Editor-in-Chief: Mark W. Huth
Associate Editor: Joan Gill

For information address Delmar Publishers, Inc.
2 Computer Drive West, Box 15-015,
Albany, New York 12212

Printed in the United States of America
Published simultaneously in Canada
by Nelson Canada,
a division of International Thomson Limited

10 9 8 7 6 5 4

Library of Congress Cataloging-in-Publication Data

Complete undercar systems.

Includes index.
1. Automobiles—Springs and suspension. 2. Automobiles
—Steering-gear. 3. Automobiles—Springs and suspension
—Maintenance and repair. 4. Automobiles—Steering-gear
—Maintenance and repair. I. Scharff, Robert.
TL257.C65 1989   629.2'43   88-33520
ISBN 0-8273-3571-7 (pbk.)
ISBN 0-8273-3573-3 (shop manual)
ISBN 0-8273-3572-5 (instructor's guide)

CONTENTS

CHAPTER 1

INTRODUCTION TO THE UNDERCAR

PRACTICE QUESTIONS

1. What component of typical front and rear suspension systems is indicated by letter A in Figure 1-1?

   a. shock absorbers
   b. control arms
   c. power steering mechanism
   d. strut

2. What component of typical front and rear suspension systems is indicated by letter B in Figure 1-1?

   a. shock absorbers
   b. control arms
   c. power steering mechanism
   d. strut

3. What steering system component is indicated by the letter A in Figure 1-2?

   a. steering column
   b. ball joints
   c. linkage
   d. power steering mechanism

4. What steering system component is indicated by the letter B in Figure 1-2?

   a. steering column
   b. ball joints
   c. linkage
   d. power steering mechanism

5. What steering system component is indicated by the letter C in Figure 1-2?

   a. steering column
   b. ball joints
   c. linkage
   d. power steering mechanism

6. On a repair order, Mechanic A writes the customer's name and the billing. Mechanic B also writes a description of the service and the parts used. Who is right?

   a. Mechanic A
   b. Mechanic B

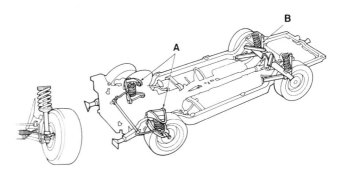

Figure 1-1

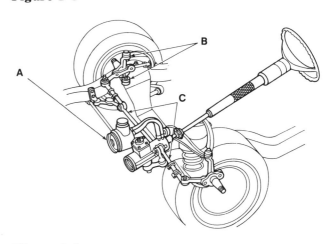

Figure 1-2

   c. Both A and B
   d. Neither A nor B

7. When diagnosing a problem with the undercar system, Mechanic A tries to find out information about the undercar system, such as its age and its service history. Mechanic B tries to find out information about the problem, such as if it started suddenly or gradually or followed a specific occurrence such as an accident. Who is right?

   a. Mechanic A
   b. Mechanic B
   c. Both A and B
   d. Neither A nor B

8. Which of the following measurements are included on an undercar analysis and alignment record?

a. caster
b. toe-in
c. camber
d. all of the above

9. What components of a front suspension are not included on the rear suspension?
a. movable wheels for steering
b. shock absorbers
c. springs
d. struts

10. A vehicle enters the shop with uneven tire wear. Mechanic A suspects problems with the brake system. Mechanic B suspects problems in steering and suspension performance. Who is right?

a. Mechanic A
b. Mechanic B
c. Both A and B
d. Neither A nor B

SHOP ASSIGNMENT 1

Examine a power steering system and locate the following components: steering column, grease fittings, ball joints, power steering mechanism, linkage, and front-wheel drive (if applicable).

JOB SHEET

SHOP ASSIGNMENT 2
RAISE AND LOWER A VEHICLE ON A LIFT

NAME _____ STATION _____ DATE _____

Tools and Materials

Lift or hoist with operating manual
Manufacturer's manual for vehicle to be used

Protective Clothing

None required

Procedure

1. Position the arms and supports of the lift to provide unobstructed clearance.

   Task completed _____

2. Load the vehicle on the lift.

   Task completed _____

   a. Are the adapters and axle supports in secure contact with the vehicle per the manual's instructions?

   Yes _____ No _____

   b. If no, correct.

   Task completed _____

3. Position the lift supports to contact at the vehicle manufacturer's recommended lifting points.

   Task completed _____

4. Securely close all doors, hood, and trunk.

   Task completed _____

5. Raise the lift until the supports contact the vehicle.

   Task completed _____

6. Lift the vehicle to the desired height.

   Task completed _____

   a. Lower the unit onto mechanical safeties.

   Task completed _____

7. To lower the vehicle, first release locking devices as per instructions.

Task completed _____

   a. Lower the lift.

Task completed _____

8. Position the arms, adapters, or axle supports out of the way.

Task completed _____

9. Remove the vehicle from the lift area.

Task completed _____

PROBLEMS ENCOUNTERED: _____

_____

_____

INSTRUCTOR'S COMMENTS: _____

_____

_____

CHAPTER 2

FRONT-SUSPENSION SYSTEMS AND THEIR COMPONENTS

PRACTICE QUESTIONS

1. What type of vehicle construction has no separate frame?
   a. conventional
   b. body-over-frame
   c. unibody
   d. fixed-axle

2. Figure 2-1 shows several designs of what type of front suspension?
   a. ball joint suspension
   b. strut type suspension
   c. parallelogram suspension
   d. SLA suspension

3. What measurement is shown in Figure 2-2?
   a. camber
   b. caster
   c. toe
   d. rebound

4. What is the most commonly used spring on traditional independent front suspensions?
   a. leaf springs
   b. coil springs
   c. torsion bar springs
   d. monoleaf springs

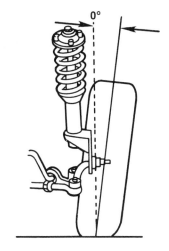

Figure 2-2

5. Which of the following is considered to be unsprung weight on a vehicle?
   a. transmission
   b. frame
   c. engine
   d. steering linkage

6. What component of a conventional shock absorber is indicated by the letter A in Figure 2-3?
   a. upper mounting
   b. reserve tube

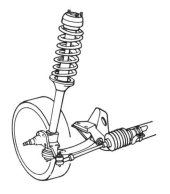
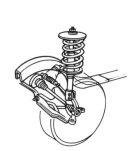
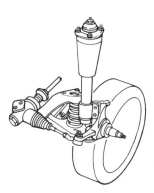

Figure 2-1

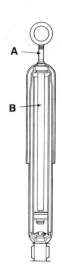

A

B

Figure 2-3

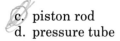

c. piston rod
d. pressure tube

7. What component of a conventional shock absorber is indicated by the letter B in Figure 2-3?

a. upper mounting
b. reserve tube
c. piston rod
d. pressure tube

8. What component of a rack-and-pinion Mac-Pherson system is indicated by the letter A in Figure 2-4?

a. lower control arm
b. strut damper
c. drive axle
d. stabilizer

9. What component of a rack-and-pinion Mac-Pherson system is indicated by the letter B in Figure 2-4?

a. lower control arm
b. strut damper
c. drive axle
d. stabilizer

10. Twin I-beam suspensions can be found on _____ .

a. two-wheel drive trucks
b. four-wheel drive trucks
c. vans
d. all of the above

SHOP
ASSIGNMENT 3

Examine a leaf spring and identify the related hardware: rear mounting shackle, spring clips, U-bolts and U-bolt fastener plate, and front mounting bolt and bushing.

SHOP
ASSIGNMENT 4

Examine a stabilizer bar on a vehicle and identify the following bar hardware: retainer, link bolt, and bushings.

SHOP
ASSIGNMENT 5

Examine a torsion bar suspension system and identify the following components: strut rod and bushing, torsion bar, stabilizer assembly, shock, upper ball joint, and lower and upper control arm bushing.

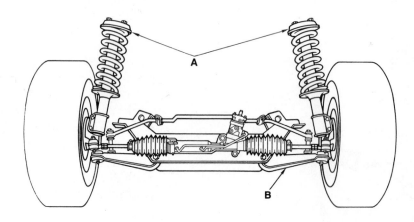

A

B

Figure 2-4

Chapter 2   Front-Suspension Systems and their Components

CHAPTER 3

FRONT-SUSPENSION SERVICING

PRACTICE QUESTIONS

1. Which of the following front-suspension components must be replaced as an assembly with other components and not as an individual component?

   a. coil spring
   b. lower arm inner pivot bushing
   c. steering knuckle
   d. forged lower arm

2. Before making a visual inspection and height measurements of a vehicle's suspension, Mechanic A makes sure the gas tank is empty. Mechanic B makes sure the gas tank is full. Who is right?

   a. Mechanic A
   b. Mechanic B
   c. Both A and B
   d. Neither A nor B

3. Mechanic A uses coil spring stabilizers to refurbish a weak spring. Mechanic B uses coil spring stabilizers to control dipping and sidesway. Who is right?

   a. Mechanic A
   b. Mechanic B
   c. Both A and B
   d. Neither A nor B

4. What tool is pictured in Figure 3-1?

   a. ball joint press
   b. spacer tool
   c. dial indicator
   d. strut compressor

5. What is the function of the tool pictured in Figure 3-1?

   a. to make a radial check
   b. to make an axial check
   c. to make precise measurements of ball joints
   d. all of the above

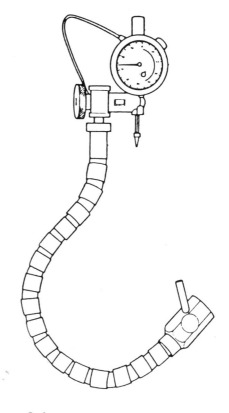

Figure 3-1

6. Which of the following is an indication of shock absorber failure?

   a. excessive bouncing after stops
   b. unusual tire wear patterns
   c. difficult steering and handling
   d. all of the above

7. Mechanic A notices a light film of oil on the upper portion of the shock absorber and considers it to be normal. Mechanic B notices the light film of oil and replaces the shock absorber. Who is right?

   a. Mechanic A
   b. Mechanic B
   c. Both A and B
   d. Neither A nor B

8. When servicing struts, Mechanic A secures the strut in a vise. Mechanic B secures the strut with a vise-holding tool. Who is right?

   a. Mechanic A
   (b.) Mechanic B
   c. Both A and B
   d. Neither A nor B

9. What component of a MacPherson strut assembly is indicated by the letter A in Figure 3-2?

   (a.) spindle
   b. strut rod bushing
   c. ball joint
   d. cartridge

10. What component of a MacPherson strut assembly is indicated by the letter B in Figure 3-2?

   a. spindle
   b. strut rod bushing
   (c.) ball joint
   d. cartridge

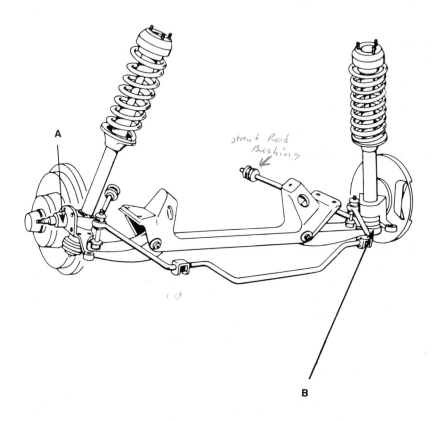

strut Rod Bushing

Figure 3-2

JOB SHEET

SHOP ASSIGNMENT 6
———— ADJUST A LONGITUDINAL TORSION BAR ————

NAME ——————————— STATION ——————————— DATE ————————

Tools and Materials

Alignment rack, optional
Torsion bar height gauge
Assorted hand tools

Protective Clothing

Safety goggles or glasses with side shields

Procedure

1. Position the car on a level surface, preferably an alignment rack.

   Task completed ————

2. Install a torsion bar gauge.

   Task completed ————

3. Compare the height of the right side to the left.

   Task completed ————

   a. Do measurements agree with alignment specifications for height tables?

   Yes ———— No ————

   b. If yes, go to step 6.

   Task completed ————

4. If height must be increased, adjust the bar clockwise.

   Not applicable ———— Task completed ————

5. If height must be decreased, adjust the bar counterclockwise.

   Not applicable ———— Task completed ————

6. Jounce the vehicle.

   Task completed ————

7. Recheck measurements on both sides.

   Task completed ————

PROBLEMS ENCOUNTERED: _____

_____

_____

INSTRUCTOR'S COMMENTS: _____

_____

_____

JOB SHEET

SHOP ASSIGNMENT 7
—— REPLACE A TWIN I-BEAM SPRING ——

NAME _____ STATION _____ DATE _____

Tools and Materials

Lift or jack stands
Floor jack
Torque wrench
Other hand tools

Protective Clothing

Safety goggles or glasses with side shields

Procedure

1. Place the vehicle frame on jack stands or a lift.

   Task completed _____

   a. Remove the tire.

   Task completed _____

   b. Support the axle below the coil spring with a floor jack.

   Task completed _____

2. Disconnect the shock absorber from the bracket.

   Task completed _____

   a. Push the shock absorber up out of the way.

   Task completed _____

3. Remove the two spring upper retainer bolts from the top of the spring seal.

   Task completed _____

   a. Remove the retainer.

   Task completed _____

4. Remove the nut attaching the spring tower retainer to the seal and I-beam.

   Task completed _____

   a. Remove the retainer.

   Task completed _____

   b. Lower the I-beam.

   Task completed _____

c. Remove the spring.

Task completed ____

5. To install the spring, place it over the lower seat.

Task completed ____

   a. Raise the I-beam.

Task completed ____

6. Torque the upper retainer attaching bolts to specifications.

Task completed ____

7. Position the lower retainer plate over the bolt.

Task completed ____

   a. Install the line attaching nut.

Task completed ____

   b. Torque to specifications.

Task completed ____

PROBLEMS ENCOUNTERED: _____

_____

_____

INSTRUCTOR'S COMMENTS: _____

_____

_____

JOB SHEET

SHOP ASSIGNMENT 8
REPLACE STRUT ROD BUSHING

NAME _____ STATION _____ DATE _____

Tools and Materials

Lift or jack stands
Torque wrench and other hand tools

Protective Clothing

Safety goggles or glasses with side shields

Procedure

1. Lift the vehicle under the frame.

Task completed _____

2. Remove the front strut rod nut.

Task completed _____

3. Install new bushings, following manufacturer's instructions.

Task completed _____

4. Hold up the rod in the rear and tighten the front nut to seat the bushings properly.

Task completed _____

5. Reconnect the rear of the strut rod to the control arm.

Task completed _____

6. Torque all nuts to specifications.

Task completed _____

7. Check the alignment.

Task completed _____

PROBLEMS ENCOUNTERED: _____

_____

INSTRUCTOR'S COMMENTS: _____

_____

REAR-SUSPENSION SYSTEMS AND THEIR SERVICE

PRACTICE QUESTIONS

1. The condition that occurs when leaf springs counteract the axle's torque reaction by flexing upward in front of the axle and downward behind it is called _____ .
   a. spring windup
   b. axle windup
   c. axle tramp
   d. wheel hop

2. What component of a link-type rigid axle suspension system is indicated by the letter A in Figure 4-1?
   a. tracking (panhard) bar
   b. control arm
   c. strut rod
   d. shock absorber

3. What component of a link-type rigid axle suspension system is indicated by the letter B in Figure 4-1?
   a. tracking (panhard) bar
   b. control arm
   c. strut rod
   d. shock absorber

4. What component of a link-type rigid axle suspension system is indicated by the letter C in Figure 4-1?
   a. tracking (panhard) bar
   b. control arm
   c. strut rod
   d. shock absorber

5. When removing the rear springs on a semi-independent suspension system, Mechanic A uses a twin-post type hoist. Mechanic B performs this operation on the floor. Who is right?
   a. Mechanic A
   b. Mechanic B

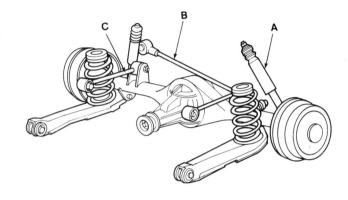

Figure 4-1

   c. Both A and B
   d. Neither A nor B

6. What is the first step in most rear suspension system servicing?
   a. Remove the rear springs.
   b. Remove both shock absorbers.
   c. Remove one shock absorber.
   d. Remove the stabilizer bar.

7. Independent suspensions can be found on _____ .
   a. only heavy-duty vehicles
   b. both FWD and RWD vehicles
   c. only FWD vehicles
   d. only RWD vehicles

8. Tight ball joints could cause _____ .
   a. front-end noise
   b. hard steering
   c. poor returnability
   d. all of the above

9. What other defective suspension component could cause similar problems as tight ball joints?
   a. dry or worn strut bearing
   b. worn or loose shocks

c. worn control arm bushings
d. all of the above

10. A vehicle in the shop pulls to one side. Mechanic A checks for worn or loose strut rod bearings. Mechanic B checks for broken or weak springs. Who is right?

    a. Mechanic A
    b. Mechanic B
    c. Both A and B
    d. Neither A nor B

SHOP
ASSIGNMENT 9

Examine an independent rear suspension and locate the following components: control arm, cross member, shock absorbers, drive half-shaft, stub axle, coil springs, and trailing arm, if applicable. Refer to Figure 4–2 as needed.

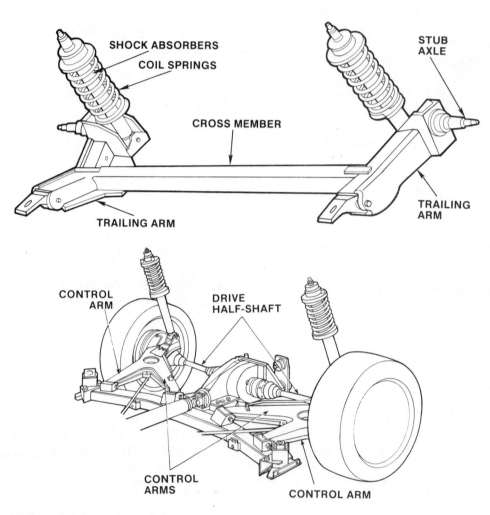

Figure 4-2

Chapter 4   Rear-Suspension Systems and their Service

JOB SHEET

SHOP ASSIGNMENT 10
REMOVE AND INSTALL A REAR SHOCK ABSORBER
FROM A LIVE-AXLE SUSPENSION SYSTEM

NAME _____ STATION _____ DATE _____

Tools and Materials

Assorted wrenches
Hoist
Adjustable jack stands (if applicable)

Protective Clothing

Safety goggles or glasses with side shields

Procedure

1. Remove the upper shock attaching nut.

Not applicable _____ Task completed _____

2. Raise the vehicle on a hoist.

Task completed _____

   a. Support the rear axle assembly.

Task completed _____

3. Remove the lower attaching bolt and nut.

Task completed _____

4. Remove the shock absorber.

Task completed _____

5. To install a shock absorber, reverse the preceding steps.

Task completed _____

PROBLEMS ENCOUNTERED: _____

_____

_____

INSTRUCTOR'S COMMENTS: _____

_____

_____

JOB SHEET

SHOP ASSIGNMENT 11
REMOVE A CONTROL ARM BUSHING FROM A
SEMI-INDEPENDENT SUSPENSION SYSTEM

NAME _____ STATION _____ DATE _____

Tools and Materials

Assorted hand tools
Bushing receiver and remover tools
Hoist and jack stands

Protective Clothing

Safety goggles or glasses with side shields

Procedure (Replace only one bushing at a time.)

1. Raise the vehicle on a hoist.

   Task completed _____

2. Remove the wheel and tire assembly.

   Task completed _____

   a. Support the body with jack stands.

   Task completed _____

3. Disconnect the brake line bracket from the body.

   Task completed _____

   a. Disconnect the parking brake cable from the hook guide on the body.

   Not applicable _____ Task completed _____

4. Remove the nut, bolt, and washer from the control arm and bracket attachment.

   Task completed _____

   a. Rotate the control arm downward.

   Task completed _____

5. Install the receiver tool on the control arm over the bushing.

   Task completed _____

   a. Tighten the attachment nuts until the tool is securely in place.

   Task completed _____

   b. Install the plate bolt through the plate and into the receiver tool.

   Task completed _____

6. Place the remover tool into position on the bushing.

Task completed _____

   a. Install the nut onto the plate bolt.

Task completed _____

   b. Remove the bushing from the control arm by turning the bolt.

Task completed _____

PROBLEMS ENCOUNTERED: _____

_____

_____

INSTRUCTOR'S COMMENTS: _____

_____

_____

JOB SHEET

SHOP ASSIGNMENT 12
INSTALL A CONTROL ARM BUSHING ON A
—— SEMI-INDEPENDENT SUSPENSION SYSTEM ——

NAME _____ STATION _____ DATE _____

Tools and Materials

Torque wrench
Assorted hand tools
Bushing receiver and installer tools
Screw type jack stand
Hoist

Protective Clothing

Safety goggles or glasses with side shields

Procedure

1. With the vehicle up on the hoist, install the bushing on the bolt.

   Task completed _____

   a. Position it onto the housing.

   Task completed _____

   b. Align the bushing installer arrow with the arrow on the receiver
      for proper indexing of the bushing.

   Task completed _____

2. Install the nut onto the bolt.

   Task completed _____

3. Press the bushing into the control arm by turning the bolt.

   Task completed _____

   a. Is the end flange flush against the face of the control arm?

   Yes _____ No _____

   b. If no, correct.

   Not applicable _____ Task completed _____

4. Use a screw type jack stand to position the control arm into the
   bracket.

   Task completed _____

   a. Install the bolt and nut. (Do not torque).

   Task completed _____

5. Install the brake line bracket to the frame.

Task completed ____

   a. Torque the screw to specifications.

Task completed ____

6. Reconnect the brake cables to the bracket.

Not applicable ____   Task completed ____

   a. Reinstall the brake cable to the hook.

Task completed ____

   b. Adjust the cable as necessary.

Task completed ____

7. Support the vehicle at curb height.

Task completed ____

   a. Tighten the control arm bolt to specifications.

Task completed ____

8. Remove the jack stands.

Task completed ____

9. Install the wheel assembly.

Task completed ____

10. Lower the vehicle from the hoist.

Task completed ____

PROBLEMS ENCOUNTERED: _____

_____

_____

INSTRUCTOR'S COMMENTS: _____

_____

_____

JOB SHEET

SHOP ASSIGNMENT 13
REMOVE STRUTS FROM AN
INDEPENDENT SUSPENSION

NAME _____ STATION _____ DATE _____

Tools and Materials

Allen wrench
Deep socket with an external hex
Safety stands or floor safety jacks

Protective Clothing

Safety goggles or glasses with side shields

Procedure

1. Loosen but do not remove the top strut attaching nut using a deep socket with an external hex, while holding the strut rod with an Allen wrench.

   Task completed _____

   a. Can the strut be reused?

   Yes _____ No _____

   b. If yes, do not grip the shock absorber shaft with pliers or vise grips to avoid damaging the shaft surface finish.

   Not applicable _____ Task completed _____

2. Raise the vehicle on a hoist.

   Task completed _____

   a. Remove the wheel and tire assembly.

   Task completed _____

   b. If using a frame contact hoist, support the lower control arm with safety stands.

   Not applicable _____ Task completed _____

   c. If using a twin post hoist, support the body with floor safety jacks on the lifting pads forward of the tie-rod body bracket.

   Not applicable _____ Task completed _____

3. Remove the clip retaining the brake flexible hose to the rear shock.

   Task completed _____

   a. Carefully move the hose aside.

   Task completed _____

4. Loosen the two nuts and bolts that retain the shock to the spindle.

Task completed _____

5. Remove the top mounting nut, washer, and rubber insulator.

Task completed _____

6. Remove the two bottom mounting bolts.

Task completed _____

    a. Remove the shock from the vehicle.

Task completed _____

PROBLEMS ENCOUNTERED: _____

_____

_____

INSTRUCTOR'S COMMENTS: _____

_____

_____

JOB SHEET

SHOP ASSIGNMENT 14
INSTALL STRUTS ON AN
INDEPENDENT SUSPENSION

NAME ———————————— STATION ———————————— DATE ——————

Tools and Materials

New upper and lower washer and insulator assemblies
New mounting bolts
Tire lubricant compatible with the rubber insulator
Allen wrench
Deep socket with an external hex

Protective Clothing

Safety goggles or glasses with side shields

Procedure

1. With the vehicle raised, extend the shock absorber to its maximum length.

   Task completed ——

2. Install a new lower washer and insulator assembly, using tire lubricant to ease insertion into the quarter panel shock tower.

   Task completed ——

3. Position the upper part of the shock absorber shaft into the shock tower opening in the body.

   Task completed ——

   a. Push slowly on the lower part of the shock until the mounting holes are lined up with mounting holes in the spindle.

   Task completed ——

4. Install new lower mounting bolts and nuts. (Do not tighten yet.)

   Task completed ——

   a. Are the heads of both bolts to the rear of the vehicle?

   Yes —— No ——

   b. If no, correct.

   Not applicable —— Task completed ——

5. Place new upper insulator and washer assembly and nut on the upper shock absorber shaft.

   Task completed ——

a. Tighten the nut to specified torque, using a deep socket with an external hex while holding the strut shaft with an Allen wrench.

Task completed _____

6. Tighten two lower mounting bolts to the specified torque.

Task completed _____

7. Install the brake flex hose.

Task completed _____

a. Install the retaining clip.

Task completed _____

8. Install the wheel and tire assembly.

Task completed _____

9. Lower the vehicle.

Task completed _____

PROBLEMS ENCOUNTERED: _____

_____

_____

INSTRUCTOR'S COMMENTS: _____

_____

_____

CHAPTER 5
ELECTRONIC AIR OR LEVEL CONTROL SYSTEMS

PRACTICE QUESTIONS

1. What component of a typical regenerative air dryer is indicated by the letter A in figure 5-1?

   a. spring
   b. dry chemical
   c. filter
   d. minimum air pressure valve

2. What component of a typical regenerative air dryer is indicated by the letter B in Figure 5-1?

   a. spring
   b. dry chemical
   c. filter
   d. minimum air pressure valve

3. What component of an exhaust solenoid is indicated by the letter A in Figure 5-2?

   a. terminals
   b. O-rings
   c. solenoid locator tabs
   d. service vents

4. What component in an EAS system continuously monitors input from the height sensors and the ignition circuits?

   a. compressor relay
   b. computer module
   c. exhaust solenoid
   d. air dryer

5. To adjust an electronic suspension system, Mechanic A works on the vehicle while it is very warm. Mechanic B allows the vehicle to warm or cool to near the temperature of the area before working on the vehicle. Who is right?

   a. Mechanic A
   b. Mechanic B
   c. Both A and B
   d. Neither A nor B

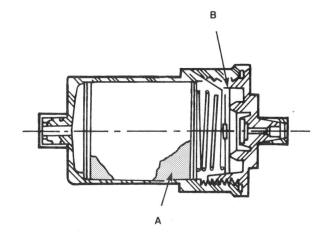

Figure 5-1

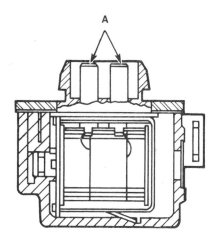

Figure 5-2

6. Which of the following measurements are affected by curb height?

   a. caster
   b. camber
   c. toe
   d. all of the above

7. Mechanic A removes an air spring when there is no pressure in it. Mechanic B removes an air spring with a little bit of pressure in it. Who is right?

a. Mechanic A
b. Mechanic B
c. Both A and B
d. Neither A nor B

8. What component in an electronic air suspension system is not present in an electronic level control system?

a. air dryer
b. compressor relay
c. computer module
d. exhaust solenoid

9. The compressor in an ELC is inoperative. Mechanic A replaces the solenoid exhaust valve assembly. Mechanic B replaces the motor cylinder assembly. Who is right?

a. Mechanic A
b. Mechanic B
c. Both A and B
d. Neither A nor B

10. What type of suspension system still in the experimental stages constantly adapts to changing road and driving conditions?

a. active suspension
b. springless electronic suspension
c. computer-controlled suspension
d. all of the above

JOB SHEET

SHOP ASSIGNMENT 15
CONDUCT A COMPLETE ADJUSTMENT PROCEDURE
OF AN ELECTRONIC SUSPENSION SYSTEM

NAME _____ STATION _____ DATE _____

Tools and Materials

Alignment rack

Protective Clothing

None required

Procedure

1. Position the vehicle on the alignment rack.

   Task completed _____

   a. Turn off the engine and exit the vehicle.

   Task completed _____

2. Level the rack as needed.

   Task completed _____

   a. Re-enter the vehicle.

   Task completed _____

   b. Turn the ignition to the RUN position but do not start it.

   Task completed _____

3. Allow the vehicle to level for 1 minute.

   Task completed _____

4. Push the luggage compartment release.

   Task completed _____

   a. Turn the ignition to the OFF position.

   Task completed _____

   b. Exit the vehicle.

   Task completed _____

5. Allow 20 seconds for the vehicle to vent to the curb height.

   Task completed _____

   a. Turn the air suspension switch to the OFF position.

   Task completed _____

6. Check the pinion angle as per instructions in the vehicle's shop service manual.

Task completed _____

PROBLEMS ENCOUNTERED: _____

_____

_____

INSTRUCTOR'S COMMENTS: _____

_____

_____

JOB SHEET

SHOP ASSIGNMENT 16
ENTER DIAGNOSTICS AND TEST DESCRIPTIONS ON AN ELECTRONIC SUSPENSION SYSTEM

NAME _____ STATION _____ DATE _____

Tools and Materials

Appropriate screwdrivers
Battery charger

Protective Clothing

None required

Procedure

1. Make sure the diagnostic pigtail is ungrounded.

   Task completed _____

   a. Turn on the suspension switch.

   Task completed _____

2. Install the battery charger.

   Task completed _____

3. With the driver's door open, cycle the ignition from the OFF to the RUN position.

   Task completed _____

   a. Keep it in the RUN position at least 5 seconds.

   Task completed _____

   b. Turn to the OFF position.

   Task completed _____

4. Ground the diagnostic pigtail by attaching a lead from the pigtail to a vehicle ground.

   Task completed _____

5. Turn the ignition switch to the RUN position. (Do not start the vehicle.)

   Task completed _____

   a. Look for the warning light to blink continuously at a rate of 1.8 blinks per second to indicate that the diagnostics have been entered.

   Task completed _____

6. Close and open the driver's door once to initiate Test 1.

Task completed _____

a. Look for the warning light to blink, pause, blink, pause, etc., until the next test is started.

Task completed _____

7. Close and open the door to cause the module to advance to the next step in the test sequence.

Task completed _____

a. Look for the warning light to blink the current test number.

Task completed _____

PROBLEMS ENCOUNTERED: _____

_____

_____

INSTRUCTOR'S COMMENTS: _____

_____

_____

JOB SHEET

SHOP ASSIGNMENT 17
CONDUCT AN OPERATIONAL CHECK
FOR A TYPICAL HEIGHT SENSOR

NAME _____ STATION _____ DATE _____

Tools and Materials

Hoist
Adjustable jacks
Appropriate screwdrivers
Appropriate wrenches

Protective Clothing

None required

Procedure

1. Turn on the ignition switch.

Task completed _____

   a. Raise the car on the hoist.

Task completed _____

   b. If a frame hoist is used, support the rear wheels and axle with adjustable jacks.

Not applicable _____ Task completed _____

   c. Adjust the jacks upward until the axle housing and/or wheels reach the curb height position.

Not applicable _____ Task completed _____

2. Compare the neutral position of the height sensor metal arm with the position of the sensor arm being tested.

Task completed _____

   a. Does the neutral arm vary more than 3 to 4 inches?

Yes _____ No _____

   b. If no, go to step 3.

Not applicable _____ Task completed _____

   c. If yes, make sure that the correct sensor mounting bolts are tight.

Not applicable _____ Task completed _____

   d. Make sure the sensor mounting bracket is not bent.

Not applicable _____ Task completed _____

e. Make any necessary corrections.

Not applicable _____ Task completed _____

3. Disconnect the link from the height sensor arm.

Task completed _____

4. Disconnect the wiring to the height sensor.

Task completed _____

   a. Reconnect the wiring to reset the sensor time limit function.

Task completed _____

5. Move the sensor metal arm upward approximately 1-1/2 to 2 inches above the neutral position.

Task completed _____

   a. Wait 8 to 15 seconds before the compressor turns on.

Task completed _____

6. As soon as the shocks noticeably inflate, move the sensor arm down slowly until the compressor stops.

Task completed _____

   a. Note the position where the compressor stops. (It should be close to the neutral position.)

Task completed _____

7. Move the arm down approximately 1-1/2 inches below the point where the compressor stopped.

Task completed _____

   a. Wait 8 to 15 seconds before the shocks start to deflate.

Task completed _____

   b. Allow the shocks to deflate until only the retention pressure is left in the shocks (approximately 8 to 15 pounds).

Task completed _____

PROBLEMS ENCOUNTERED: _____

_____

_____

INSTRUCTOR'S COMMENTS: _____

_____

_____

JOB SHEET

SHOP ASSIGNMENT 18
CONDUCT A COMPONENT
PERFORMANCE TEST

NAME _____ STATION _____ DATE _____

Tools and Materials

Ammeter
Assorted hand tools
Service manual

Protective Clothing

Safety goggles or glasses with side shields

Procedure

1. Disconnect the wiring from the compressor motor.

Task completed _____

   a. Disconnect the wiring from the exhaust solenoid terminals.

Task completed _____

2. Disconnect the existing pressure line from the dryer.

Task completed _____

   a. Attach the pressure gauge to the dryer fitting.

Task completed _____

3. Connect the ammeter to a 12-volt source and to the compressor. (Current draw should not exceed 14 amperes.)

Task completed _____

4. When the gauge reads 110 to 120 psi, shut off the compressor.

Task completed _____

   a. Does the pressure leak down? (If the compressor runs until it reaches its maximum output pressure, the solenoid exhaust valve will act as a relief valve, and the resulting leak down when the compressor is shut off will indicate a false leak.)

Yes _____ No _____

5. Does leak-down pressure drop below 90 psi when the compressor is shut off?

Yes _____ No _____

   a. If yes, use a diagnostic chart in a service manual to solve the problem.

Not applicable _____ Task completed _____

PROBLEMS ENCOUNTERED: _____

_____

_____

INSTRUCTOR'S COMMENTS: _____

_____

_____

CHAPTER 6

MANUAL STEERING COMPONENTS AND SERVICE

PRACTICE QUESTIONS

1. What component of a conventional manual steering system is indicated by the letter A in Figure 6–1?

   a. steering wheel
   b. steering shaft
   c. steering linkage
   d. gearbox

2. What component of a conventional manual steering system is indicated by the letter B in Figure 6–1?

   a. steering wheel
   b. steering shaft
   c. steering linkage
   d. gearbox

3. What component in parallelogram steering linkage makes the final connections between the steering linkage and steering knuckles?

   a. tie-rod assembly
   b. center link
   c. idler arm
   d. pitman arm

4. A vehicle enters the shop with road shimmy. Mechanic A blames the manual steering gear. Mechanic B blames the steering wheel. Who is right?

   a. Mechanic A
   b. Mechanic B
   c. Both A and B
   d. Neither A nor B

5. A vehicle enters the shop with excessive play or looseness in the steering system. Mechanic A looks for worn steering shaft couplings or worn upper ball joints. Mechanic B looks for loosely adjusted steering gear thrust bearings or front-wheel bearings. Who is right?

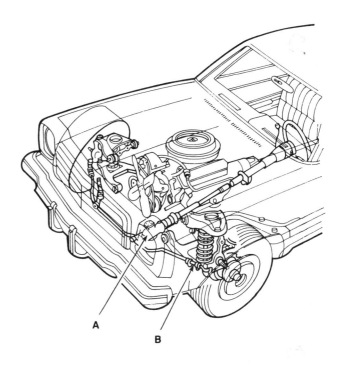

Figure 6-1

   a. Mechanic A
   b. Mechanic B
   c. Both A and B
   d. Neither A nor B

6. Which of the following conditions cannot be caused by manual steering gear?

   a. vehicle pulls or drifts to one side
   b. shimmy
   c. both a and b
   d. neither a nor b

7. A vehicle enters the shop with noise in the automatic transmission steering column. Mechanic A checks the couplings and column alignment. Mechanic B checks bearings and the shaft lock plate. Who is right?

   a. Mechanic A
   b. Mechanic B

c. Both A and B
d. Neither A nor B

8. The steering wheel fails to return to the top tilt position in a vehicle with a manual transmission column. Mechanic A says the shoe is seized on its pivot pin. Mechanic B says the wheel tilt spring is defective. Who is right?
   a. Mechanic A
   b. Mechanic B
   c. Both A and B
   d. Neither A nor B

9. Before removing a steering column, Mechanic A sets the parking brake. Mechanic B removes the battery cable from the negative terminal. Who is right?
   a. Mechanic A
   b. Mechanic B
   c. Both A and B
   d. Neither A nor B

10. What component involved in a typical tie-rod end position is indicated by the letter A in Figure 6-2?
    a. steering knuckle
    b. adjusting sleeve
    c. idler arm
    d. pitman arm

SHOP ASSIGNMENT 19

Examine a typical parallelogram steering linkage and locate the following components placed behind the front suspension: center link, sockets, pitman arm, tie-rods, idler arm, and frame bracket. Refer to Figure 6-3 as needed.

SHOP ASSIGNMENT 20

Examine a typical rack-and-pinion steering system and locate the following components: steering gear, tie-rods, outer tie-rod end, bellows, and inner socket assembly. Refer to Figure 6-4 as needed.

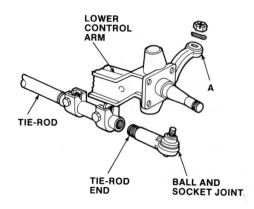

Figure 6-2

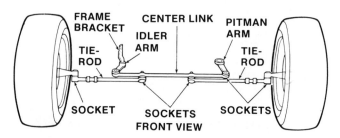

Figure 6-3

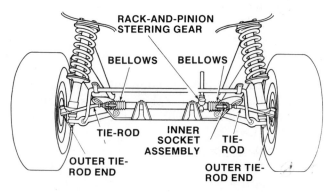

Figure 6-4

JOB SHEET

SHOP ASSIGNMENT 21
REPLACE A CENTER LINK IN PARALLELOGRAM
STEERING LINKAGE

NAME _____ STATION _____ DATE _____

Tools and Materials

Hand wrenches
Torque wrench

Protective Clothing

Safety goggles or glasses with side shields

Procedure

1. Remove the cotter pins and nuts at all connections.

Task completed _____

2. Break the taper between the joints.

Task completed _____

   a. Remove the center link.

Task completed _____

3. Clean and inspect the mounting holes.

Task completed _____

   a. Clean and inspect the tapered studs on the tie-rods.

Task completed _____

   b. Clean and inspect the idler arm.

Task completed _____

   c. Clean and inspect the pitman arm.

Task completed _____

4. Are any holes or studs worn?

Yes _____ No _____

   a. If yes, replace.

Not applicable _____ Task completed _____

5. Install the new center link.

Task completed _____

   a. Tighten the nuts and torque to specifications.

Task completed _____

6. Install the cotter pins.

Task completed _____

7. Check alignment for correct toe angle.

Task completed _____

PROBLEMS ENCOUNTERED: _____

_____

_____

INSTRUCTOR'S COMMENTS: _____

_____

_____

JOB SHEET

SHOP ASSIGNMENT 22
REMOVE A TYPICAL
STEERING WHEEL

NAME _____ STATION _____ DATE _____

Tools and Materials

Hand wrenches
Steering wheel remover or puller

Protective Clothing

Safety goggles or glasses with side shields

Procedure

1. Remove the steering wheel cover.

    Task completed _____

2. Loosen the steering wheel bolt four to six turns, but do not remove it.

    Task completed _____

3. Use the steering wheel remover or puller on top of the bolt until the steering wheel is loose from the shaft.

    Task completed _____

4. Remove and discard the steering wheel attaching bolt.

    Task completed _____

PROBLEMS ENCOUNTERED: _____

_____

_____

INSTRUCTOR'S COMMENTS: _____

_____

_____

JOB SHEET

SHOP ASSIGNMENT 23
INSTALL A TYPICAL
STEERING WHEEL

NAME _____ STATION _____ DATE _____

Tools and Materials

Torque wrench and other hand wrenches
New steering wheel bolt

Protective Clothing

Safety goggles or glasses with side shields

Procedure

1. Position the steering wheel on the end of the steering wheel shaft.

   Task completed _____

   a. Align the index mark on the steering wheel with the index mark on the shaft to ensure that the straight-ahead steering wheel position corresponds to the straight-ahead position of the front wheels.

   Task completed _____

2. Install a new steering wheel bolt.

   Task completed _____

   a. Tighten to manufacturer's specifications.

   Task completed _____

3. Install the steering wheel cover.

   Task completed _____

4. Check the steering column for proper operations.

   Task completed _____

PROBLEMS ENCOUNTERED: _____

_____

_____

INSTRUCTOR'S COMMENTS: _____

_____

_____

CHAPTER 7

POWER STEERING COMPONENTS AND SERVICE

PRACTICE QUESTIONS

1. What component of a typical integral power steering system is indicated by the letter A in Figure 7-1?

   a. return hose
   b. gear
   c. pressure hose
   d. pump

2. What component of a typical integral power steering system is indicated by the letter B in Figure 7-1?

   a. return hose
   b. gear
   c. pressure hose
   d. pump

3. What component of a typical linkage power steering system is indicated by the letter A in Figure 7-2?

   a. pump
   b. power cylinder
   c. center link
   d. piston rod

4. What component of a typical remote reservoir power steering system is indicated by the letter A in Figure 7-3?

   a. suction line
   b. rack-and-pinion gear
   c. remote reservoir
   d. pump

5. What component of a typical remote reservoir power steering system is indicated by the letter B in Figure 7-3?

   a. suction line
   b. rack-and-pinion gear
   c. remote reservoir
   d. pump

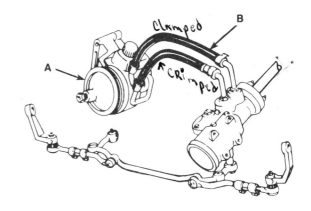

Figure 7-1

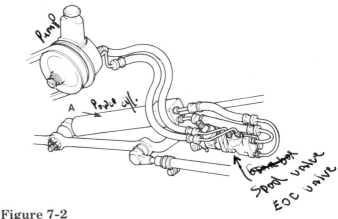

Figure 7-2

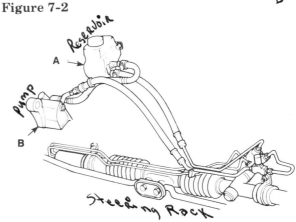

Figure 7-3

45

6. Mechanic A checks the fluid level in the reservoir with the engine cold. Mechanic B checks the fluid level in the reservoir after the engine has been run at idle for a few minutes. Who is right?
   a. Mechanic A
   b. Mechanic B
   c. Both A and B
   d. Neither A nor B

7. Which of the following vehicle components that are not part of the power steering system can affect steering performance?
   a. tires
   b. wheel bearings
   c. suspension pivots
   d. all of the above

8. A vehicle has a momentary increase in steering wheel effort when turned rapidly. Mechanic A checks the fluid level in the reservoir. Mechanic B checks for front end misalignment. Who is right?
   a. Mechanic A
   b. Mechanic B
   c. Both A and B
   d. Neither A nor B

9. In an electronically driven power rack-and-pinion system, what component replaces the hydraulic pump of the conventional system as a power source?
   a. varible assist
   b. TRW
   c. alternator
   d. transducer

10. Fluid level is checked at the _____.
   a. pump reservoir
   b. steering wheel
   c. master cylinder
   d. power cylinder

JOB SHEET

SHOP ASSIGNMENT 24
BLEED A HYDRO-BOOST
POWER STEERING SYSTEM

NAME _____ STATION _____ DATE _____

Tools and Materials

Clean power steering fluid

Protective Clothing

Safety goggles or glasses with side shields

Procedure

1. Turn on the engine.

   Task completed _____

   a. With the engine running, apply the brakes with a pumping action. (Do not turn the steering wheel until all the air is expelled from the hydro-boost valve.)

   Task completed _____

2. Stop the engine.

   Task completed _____

   a. Remove the reservoir cap.

   Task completed _____

3. Is the fluid level low?

   Yes _____ No _____

   a. If yes, fill.

   Not applicable _____ Task completed _____

4. Start the engine.

   Task completed _____

5. Turn the steering wheel from side to side several times, but avoid hitting stops.

   Task completed _____

   a. Roll the vehicle to change tire-to-floor contact area.

   Task completed _____

6. Stop the engine.

   Task completed _____

   a. Is the fluid level low?

   Yes _____ No _____

b. If yes, fill.

Not applicable _____   Task completed _____

7. Start the engine.

Task completed _____

8. Depress the brake pedal several times while rotating the steering wheel from stop to stop.

Task completed _____

9. Stop the engine.

Task completed _____

10. Pump the brake pedal four to five times to deplete the accumulator pressure.

Task completed _____

a. Is the fluid level low?

Yes _____   No _____

b. If yes, fill.

Not applicable _____   Task completed _____

11. Start the engine.

Task completed _____

a. Roll the vehicle to change tire-to-floor contact area.

Task completed _____

12. Repeat steps 1 through 7 until all air is expelled from the system with fluid at HOT level.

Task completed _____

13. Return the wheels to the center position.

Task completed _____

a. Keep engine running another 2 to 3 minutes.

Task completed _____

b. Check for leaks.

Task completed _____

14. Turn off the engine.

Task completed _____

a. Replace the reservoir cap.

Task completed _____

PROBLEMS ENCOUNTERED: _____

_____

_____

INSTRUCTOR'S COMMENTS: _____

_____

_____

JOB SHEET

SHOP ASSIGNMENT 25
REPLACE A
POWER STEERING HOSE

NAME _____ STATION _____ DATE _____

Tools and Materials

Clean power steering fluid
New power steering hose

Protective Clothing

Safety goggles or glasses with side shields

Procedure

1. Check the routing of the old hose while still in the vehicle.

    Task completed _____

    a. Lay the replacement in a place beside old hose.

    Task completed _____

2. Remove damaged hose.

    Task completed _____

3. Install new hose.

    Task completed _____

4. Tighten to specifications.

    Task completed _____

    a. Is leakage evident?

    Yes _____ No _____

    b. If no, go to step 5.

    Not applicable _____ Task completed _____

    c. If yes, loosen the nut.

    Task completed _____

    d. Rotate the hose to reseat it with the brass seat.

    Task completed _____

5. Fill the reservoir.

    Task completed _____

6. Start the engine.

    Task completed _____

a. Let it run a few seconds, then shut it off.

Task completed _____

7. Recheck the fluid level.

Task completed _____

8. Restart the engine.

Task completed _____

a. Turn the steering wheel right and left, gently contacting the stops.

Task completed _____

9. Recheck the fluid level.

Task completed _____

a. Is the fluid extremely foamy?

Yes _____ No _____

b. If no, go to the end of the page.

Not applicable _____ Task completed _____

c. If yes, let the vehicle sit for awhile.

Not applicable _____ Task completed _____

d. Repeat the above procedure.

Not applicable _____ Task completed _____

PROBLEMS ENCOUNTERED: _____

_____

_____

INSTRUCTOR'S COMMENTS: _____

_____

_____

DRIVE SHAFT AND UNIVERSAL JOINT SERVICE

PRACTICE QUESTIONS

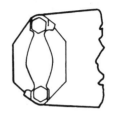

Figure 8-1

1. What type of retaining method for U-joints is depicted in Figure 8-1?

   a. thrust plate
   b. strap
   c. bearing plate
   d. U-bolt

2. What type of retaining method for U-joints is depicted in Figure 8-2?

   a. thrust plate
   b. strap
   c. bearing plate
   d. U-bolt

Figure 8-2

3. What measurement is indicated by the arrow in Figure 8-3?

   a. inside yoke span
   b. outside yoke span
   c. canceling angle
   d. centerline

4. What tool is used to check drive shaft runout?

   a. inclinometer
   b. dial indicator
   c. arbor press
   d. none of the above

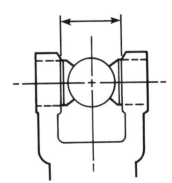

Figure 8-3

5. A vehicle has vibration in the drive shaft. Mechanic A checks for a worn transmission extension housing bushing. Mechanic B checks for loose or worn U-joints. Who is right?

   a. Mechanic A
   b. Mechanic B
   c. Both A and B
   d. Neither A nor B

6. A vehicle has a knock or clunking noise in the driveline at speeds lower than 10 mph. Mechanic A checks the differential. Mechanic B checks the U-joints. Who is right?

   a. Mechanic A
   b. Mechanic B
   c. Both A and B
   d. Neither A nor B

7. What tool is depicted in Figure 8-4?

   a. inclinometer
   b. dial indicator
   c. wheel balancer
   d. Wittek hose clamp

8. What is the use of the tool in Figure 8-4?

   a. to determine centerline
   b. to check drive shaft and U-joint angles
   c. to check drive shaft runout
   d. to balance the U-joints

9. To save time and expense when replacing U-joint parts, Mechanic A uses some old parts and some new ones. Mechanic B always replaces all parts as a complete repair kit. Who is right?

   a. Mechanic A
   b. Mechanic B
   c. Both A and B
   d. Neither A nor B

10. A condition caused by the trunnion bearing surface being broken by the U-joint's needle bearings is called _____ .

a. press-fit
b. ellipse
c. fluctuation
d. brinelling

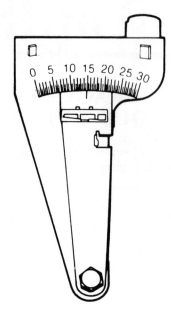

Figure 8-4

Chapter 8   Drive Shaft and Universal Joint Service

JOB SHEET

SHOP ASSIGNMENT 26
REMOVE A TYPICAL
REAR-WHEEL DRIVE SHAFT

NAME ———————————— STATION ———————————— DATE ——————

Tools and Materials

Drain pan
Hammer
Center punch
Box-end wrench
Large screwdriver or small pry bar
Masking tape

Protective Clothing

Safety goggles or glasses with side shields

Procedure

1. Raise the vehicle on a lift or hoist.

   Task completed ——

   a. Make sure it is secure.

   Task completed ——

2. Place a drain pan below the transmission output shaft to catch any
   fluid that might come out when the slip yoke is removed.

   Task completed ——

3. Are arrows located on the driving and driven parts of the drive shaft
   to maintain alignment?

   Yes —— No ——

   a. If yes, go to step 4.

   Not applicable —— Task completed ——

   b. If no, mark both the drive shaft yoke and differential pinion flange
      yoke with a hammer and center punch.

   Task completed ——

4. Use the proper size box-end wrench to remove the four attaching bolts
   that hold the U-joint retaining straps to the pinion yoke.

   Not applicable —— Task completed ——

5. Remove the hold-down straps.

   Not applicable —— Task completed ——

6. If the vehicle uses two U-bolts to retain the U-joint to the flange, remove the four U-bolt nuts and lock washers.

Not applicable _____ Task completed _____

a. Remove the U-bolts.

Not applicable _____ Task completed _____

7. Use a large screwdriver or small pry bar to separate the drive shaft yoke and trunnion from the pinion yoke.

Task completed _____

a. Carefully lower the end of the drive shaft.

Task completed _____

8. Use masking tape to secure the trunnion caps to the trunnion to prevent loss of bearing parts.

Task completed _____

9. Pull the drive shaft straight back to allow the splined slip yoke on the transmission end of the drive shaft to come free of the transmission output shaft.

Task completed _____

PROBLEMS ENCOUNTERED: _____

_____

_____

INSTRUCTOR'S COMMENTS: _____

_____

_____

JOB SHEET

SHOP ASSIGNMENT 27
DISASSEMBLE AN INSIDE
SNAP RING U-JOINT

NAME _____ STATION _____ DATE _____

Tools and Materials

Drift pin
Screwdriver
Arbor press
Tube stock

Protective Clothing

Safety goggles or glasses with side shields

Procedure

1. With the drive shaft on a bench, locate the snap rings.

   Task completed _____

   a. Tap them out with a drift pin.

   Task completed _____

   b. Does a ring spin in the groove?

   Yes _____ No _____

   c. If yes, hold one end of the ring with a screwdriver while using a drift on the other.

   Not applicable _____ Task completed _____

2. Set the shaft joint in the arbor press with a piece of tube stock beneath it. (See Figure 8-5.)

   Task completed _____

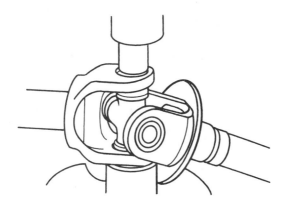

Figure 8-5

Chapter 8 Drive Shaft and Universal Joint Service   57

a. Does the cross have a zerk?

Yes _____ No _____

b. If yes, position the joint so the zerk is pointing up.

Not applicable _____ Task completed _____

3. Press.

Task completed _____

4. Does the bearing cap pull out by hand after pressing?

Yes _____ No _____

a. If no, strike the base of the lug near the cap to dislodge it.

Not applicable _____ Task completed _____

5. To remove the opposite cap, first turn the joint over.

Task completed _____

a. Straighten the cross in the open bearing hole.

Task completed _____

b. Carefully press on the trunnion so the remaining cap moves straight out of the bearing hole.

Task completed _____

6. Repeat this procedure on the remaining caps to remove the cross from the yoke.

Task completed _____

PROBLEMS ENCOUNTERED: _____

_____

_____

INSTRUCTOR'S COMMENTS: _____

_____

_____

JOB SHEET

SHOP ASSIGNMENT 28
REASSEMBLE AN INSIDE
SNAP RING U-JOINT

NAME _____ STATION _____ DATE _____

Tools and Materials

No. 2 grade EP (extreme pressure) lubricant
Light hammer, if necessary
Drift
Hand tools
Arbor press

Protective Clothing

Safety goggles or glasses with side shields

Procedure

1. During reassembly, fill the cavity in each trunnion completely with
   No. 2 EP lubricant. (Do not pack bearing caps full because this can
   burst the lubed-for-life seals.)

   Task completed _____

2. Insert the cross into the yokes; do not use force.

   Task completed _____

3. Make sure that all needle bearings are lined up inside the bearing cap.

   Task completed _____

4. Assemble the cross and bearing caps in the short shaft.

   Task completed _____

5. Assemble the other yoke.

   Task completed _____

6. When installing the last bearing caps, position the end of the trunnion flush with the outside of the bearing hole to allow the trunnion
   to act as a pilot during the press and help keep all the needle bearings
   in place.

   Task completed _____

7. Press the last caps and push slightly past the surface of the yoke so
   the snap ring will fit easily into the snap ring grooves.

   Task completed _____

   a. Inspect all four snap rings carefully.

   Task completed _____

b. Seat them with a drift.

Task completed _____

8. After reassembly, flex the joint.

Task completed _____

a. Does the joint have more than moderate stiffness?

Yes _____ No _____

b. If yes, strike the base of the lugs with a light hammer to align the
needle bearings and free up the stiff joint. (See Figure 8-6.)

Not applicable _____ Task completed _____

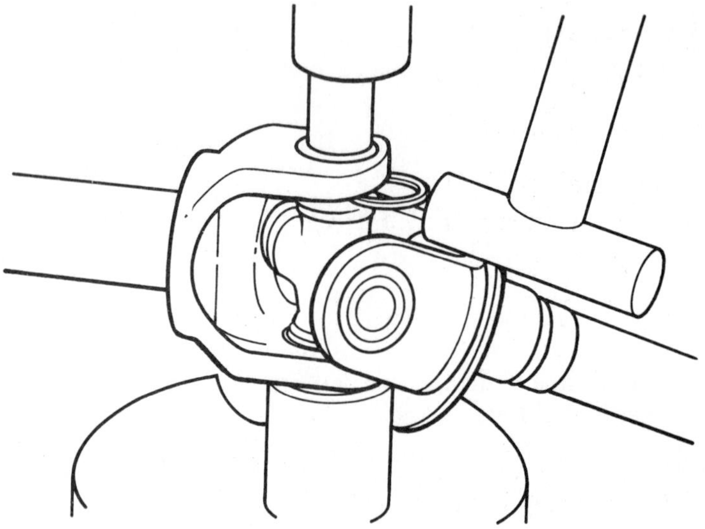

Figure 8-6

PROBLEMS ENCOUNTERED: _____

_____

_____

INSTRUCTOR'S COMMENTS: _____

_____

_____

JOB SHEET

SHOP ASSIGNMENT 29
INSTALL A TYPICAL
REAR-WHEEL DRIVE SHAFT

NAME _____ STATION _____ DATE _____

Tools and Materials

Chassis grease
Box-end wrench
Other hand tools, if needed

Protective Clothing

Safety goggles or glasses with side shields

Procedure

1. Put a small quantity of chassis grease in each bearing cup.

   Task completed _____

2. Position the slip yoke on the transmission output shaft splines.

   Task completed _____

   a. Turn the shaft until the splines are engaged. (See Figure 8-7.)

   Task completed _____

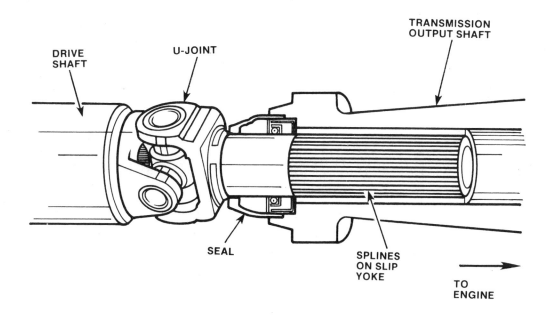

Figure 8-7

3. Push forward on the shaft and continue inching the shaft forward until the yoke is seated firmly on the output shaft.

Task completed ____

4. Toward the rear of the vehicle, line up the center punch marks or arrows on the pinion flange and the drive shaft yoke.

Task completed ____

5. Connect the rear U-joint to the pinion flange yoke.

Task completed ____

   a. Make sure the bearing cups are completely seated in the flange.

Task completed ____

6. Install the U-bolts or retaining straps.

Task completed ____

7. Install the retaining nuts or bolts.

Task completed ____

   a. Tighten them securely with the proper size box-end wrench.

Task completed ____

8. Lower the vehicle.

Task completed ____

   a. Remove it from the lift or hoist.

Task completed ____

   b. Road-test it.

Task completed ____

PROBLEMS ENCOUNTERED: _____

_____

_____

INSTRUCTOR'S COMMENTS: _____

_____

_____

JOB SHEET

SHOP ASSIGNMENT 30
REMOVE AN EXTENSION
HOUSING BUSHING*

NAME _____ STATION _____ DATE _____

Tools and Materials

Center punch
Hammer
Masking tape
Transmission jack
Clean towel
Drain pan
Bench vise
Metal chisel or screwdriver

Protective Clothing

Safety goggles or glasses with side shields

Procedure

1. Use a center punch and hammer to index the rear U-joint yoke and drive pinion flange.

   Task completed _____

2. Remove the drive shaft from the vehicle and from the immediate work area.

   Task completed _____

3. Wrap masking tape all around the rear U-joint.

   Task completed _____

4. Position a transmission jack directly beneath the transmission oil pan.

   a. Raise the jack to support the weight of the transmission.

   Task completed _____

   Task completed _____

5. Remove the cross member bolts.

   Task completed _____

   a. Remove the transmission mount bolts.

   Task completed _____

   b. Remove the cross member from the vehicle and from the immediate work area.

   Task completed _____

*If the extension housing cannot be removed from the transmission on the vehicle, follow Shop Assignment 32.

6. Disconnect the speedometer pinion assembly from the extension housing.

Task completed _____

a. Wrap it in a towel.

Task completed _____

b. Move it out of the work area.

Task completed _____

7. Position a drain pan beneath the transmission to catch any lubricant.

Task completed _____

8. Remove the transmission extension housing bolts.

Task completed _____

a. Remove the housing.

Task completed _____

b. Place the extension housing in a bench vise.

Task completed _____

9. Use a metal chisel or screwdriver to pry the extension housing seal from the extension housing.

Task completed _____

10. Find the notch in the extension housing.

Task completed _____

a. Use a chisel or screwdriver and a hammer to drive the worn bushing from the extension housing.

Task completed _____

PROBLEMS ENCOUNTERED: _____

_____

_____

INSTRUCTOR'S COMMENTS: _____

_____

_____

JOB SHEET

SHOP ASSIGNMENT 31
INSTALL AN
EXTENSION HOUSING BUSHING

NAME _____ STATION _____ DATE _____

Tools and Materials

Hammer
Bushing driver and handle
Nonhardening seal
Appropriate wrenches
Screwdriver
Lubricant
Seal driver
Hoist or insulation jacks

Protective Clothing

Safety goggles or glasses with side shields

Procedure

1. Stand the extension housing in a vertical position.

   Task completed _____

2. Align a new bushing in the housing.

   Task completed _____

   a. Use a hammer and bushing driver and handle to drive it into place.

   Task completed _____

3. Place nonhardening sealer around the outside of the seal case.

   Task completed _____

4. Install a new extension-to-transmission housing gasket or seal ring.

   Task completed _____

   a. Place the housing seal driver.

   Task completed _____

5. Align the seal with the extension housing.

   Task completed _____

   a. Drive the seal until it is seated in place.

   Task completed _____

6. Install the cross member and bolts.

   Task completed _____

7. Install the transmission mount and bolts.

Task completed _____

8. Reconnect the speedometer cable into place on the housing.

Task completed _____

9. Lubricate the outside diameter of the sliding yoke so the housing bushing and seal are lubricated for initial operation.

Task completed _____

10. Align the center punch marks.

Task completed _____

   a. Install the rear U-joint retainers.

Task completed _____

11. Lower the vehicle to the floor.

Task completed _____

   a. Operate the transmission through the various gears, checking for a clunking sound.

Task completed _____

PROBLEMS ENCOUNTERED: _____

_____

_____

INSTRUCTOR'S COMMENTS: _____

_____

_____

JOB SHEET

SHOP ASSIGNMENT 32
REMOVE AND INSTALL AN
EXTENSION HOUSING BUSHING*

NAME _____ STATION _____ DATE _____

Tools and Materials

Center punch
Hammer
Masking tape
Drain pan
Extension seal remover
Extension bushing remover
Bushing driver
Seal installer
Hoist or insulation jacks

Protective Clothing

Safety goggles or glasses with side shields

Procedure

1. Use a center punch and hammer to index the rear U-joint yoke and drive pinion flange.

   Task completed _____

2. Remove the drive shaft from the vehicle and from the immediate work area.

   Task completed _____

3. Wrap masking tape all around the rear U-joint.

   Task completed _____

4. Position a drain pan beneath the transmission to catch any lubricant.

   Task completed _____

5. Use the specially designed extension seal remover to remove the extension housing seal. (See Figure 8-8.)

   Task completed _____

6. Use the extension bushing remover (see Figure 8-8) to remove the bushing.

   Task completed _____

7. To install the housing bushing, use an extension housing bushing driver.

   Task completed _____

*This shop assignment is performed on a vehicle in which the extension housing cannot be removed from the transmission.

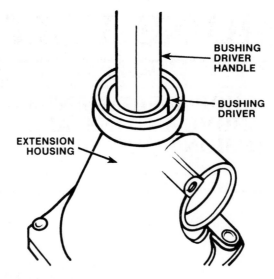

Figure 8-8

8. Use a seal installer to install the seal.

Task completed _____

9. Lubricate the outside diameter of the sliding yoke so the housing bushing and seal are lubricated for initial operation.

Task completed _____

10. Align the center punch marks.

Task completed _____

    a. Install the rear U-joint retainers.

Task completed _____

11. Lower the vehicle to the floor.

Task completed _____

    a. Operate the transmission through the various gears, checking for a clunking noise.

Task completed _____

PROBLEMS ENCOUNTERED: _____

_____

_____

INSTRUCTOR'S COMMENTS: _____

_____

_____

CHAPTER 9

CONSTANT VELOCITY JOINT SERVICE

PRACTICE QUESTIONS

1. What measurement is indicated by the arrows in Figure 9-1?

   a. camber
   b. toe
   c. caster
   d. scrub radius

2. What type of caster is depicted in Figure 9-2A?

   a. zero caster
   b. positive caster
   c. negative caster
   d. no caster

3. What type of caster is depicted in Figure 9-2B?

   a. zero caster
   b. positive caster
   c. negative caster
   d. no caster

4. What measurement is indicated by the arrows at the bottom of Figure 9-3?

   a. caster
   b. toe
   c. scrub radius
   d. included angle

5. A vehicle owner complains of steering wander or pull. Mechanic A looks for incorrect camber. Mechanic B looks for unequal caster. Who is right?

   a. Mechanic A
   b. Mechanic B
   c. Both A and B
   d. Neither A nor B

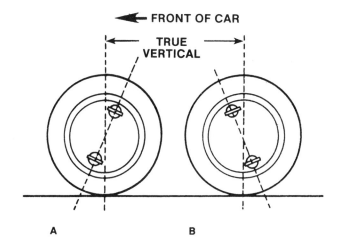

Figure 9-2

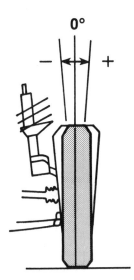

Figure 9-1

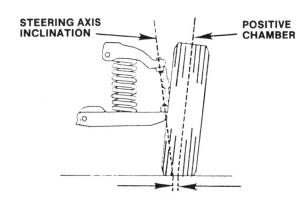

Figure 9-3

6. Before making any alignment adjustments, which of the following should be checked?

 a. Check tires for correct inflation.
 b. Check wheel balance.
 c. Check defective shock absorbers.
 d. All of the above

7. Which of the following methods or components are used to adjust caster and camber?

 a. eccentrics
 b. shims
 c. slotted frame
 d. all of the above

8. When adjusting caster and/or camber on a vehicle with a MacPherson suspension, Mechanic A replaces either the strut assembly or the control arm. Mechanic B bends the strut. Who is right?

 a. Mechanic A
 b. Mechanic B

 c. Both A and B
 d. Neither A nor B

9. Figure 9-4 shows a vehicle that is

 _____ .

 a. improperly tracking
 b. properly tracking
 c. dog tracking
 d. tracking off-center

10. When adjusting caster on a light truck with twin I-beam suspension, Mechanic A adjusts it on the right-hand axle to avoid causing a change in drive shaft alignment. Mechanic B adjusts caster on the left-hand axle. Who is right?

 a. Mechanic A
 b. Mechanic B
 c. Both A and B
 d. Neither A nor B

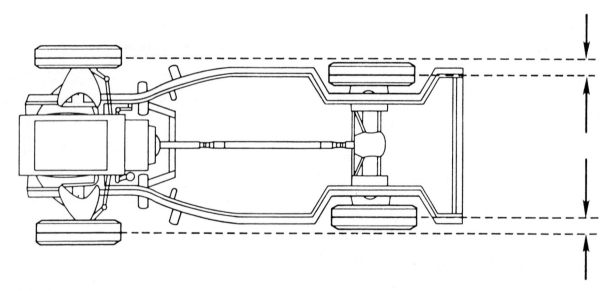

Figure 9-4

JOB SHEET

SHOP ASSIGNMENT 33
ADJUST CAMBER ON
REAR SUSPENSION

NAME _____ STATION _____ DATE _____

Tools and Materials

Alignment rack or other alignment equipment
Wrenches and other hand tools
Jack stands
New grease seals, wheel bearings, and CV or U-joints, if necessary

Protective Clothing

Safety goggles or glasses with side shields

Procedure

1. Check and record the existing camber and caster on each side to determine necessary correction.

Camber: left side _____ right side _____

Caster: left side _____ right side _____

Task completed _____

2. Support the vehicle weight under the axle tubes.

Task completed _____

3. Remove the wheel.

Task completed _____

   a. Remove the brakes.

Task completed _____

4. Remove the hub lock or original hub drive gear.

Task completed _____

   a. Remove the internal locknut.

Task completed _____

   b. Remove the bearing adjusting lock ring and nut.

Task completed _____

5. Pull the hub and rotor assembly, as applicable.

Not applicable _____ Task completed _____

   a. Pull the brake drum assembly, as applicable.

Not applicable _____ Task completed _____

6. Remove the brake backing plate.

Task completed _____

7. Remove the spindle.

Task completed _____

   a. Check the wheel spindle and bearings, and CV and U-joints.

Task completed _____

8. Place the appropriate shim on the steering knuckle studs face out on the spindle.

Task completed _____

   a. Reduce camber with the thin portion of the shim on the top of the knuckle.

Not applicable _____ Task completed _____

   b. Increase camber with the thin portion of the shim down.

Not applicable _____ Task completed _____

9. Replace all grease seals.

Not applicable _____ Task completed _____

   a. Replace worn wheel bearings.

Not applicable _____ Task completed _____

   b. Replace worn CV and U-joints.

Not applicable _____ Task completed _____

10. Reassemble.

Task completed _____

11. Remove jack stands.

Task completed _____

12. Check the new camber setting.

Task completed _____

13. Repeat procedure for the other side.

Task completed _____

PROBLEMS ENCOUNTERED: _____

_____

_____

INSTRUCTOR'S COMMENTS: _____

_____

_____

JOB SHEET

SHOP ASSIGNMENT 34
ADJUST CASTER ON A 4 × 4 TRUCK WITH LEAF SPRING FRONT SUSPENSION

NAME ————————————— STATION ————————————— DATE ——————————

Tools and Materials

Alignment rack or other alignment equipment
Safety stands
Torque wrench and other hand tools

Protective Clothing

Safety goggles or glasses with side shields

Procedure

1. Use alignment equipment to determine necessary caster adjustment.

   Task completed ——

2. Raise the vehicle.

   Task completed ——

   a. Support front axles on safety stands.

   Task completed ——

3. Loosen the U-bolt nuts on the right-hand axle.

   Task completed ——

   a. Separate the spring from the axle.

   Task completed ——

   b. Insert the shim of the required angle between the spring and axle with the thin edge of the shim toward the front of the vehicle to increase caster.

   Not applicable —— Task completed ——

   c. Insert the shim with the thin edge toward the rear to decrease caster.

   Not applicable —— Task completed ——

4. Tighten the U-bolts until they contact the cap.

   Task completed ——

   a. Torque them to specifications provided in the shop manual.

   Task completed ——

5. Lower the vehicle.

   Task completed ——

    a. Check caster again.

Task completed _____

6. If necessary, repeat the procedure on the left side.

Not applicable _____ Task completed _____

    a. If adjustments are made on the left side, check drive shaft alignment.

Not applicable _____ Task completed _____

    b. If drive shaft alignment should happen to be altered, correct it as directed in the light truck shop manual.

Not applicable _____ Task completed _____

PROBLEMS ENCOUNTERED: _____

_____

_____

INSTRUCTOR'S COMMENTS: _____

_____

_____

SEAL AND BEARING SERVICING

PRACTICE QUESTIONS

1. What component of a typical ball bearing is indicated by the letter A in Figure 10-1?

    a. inner race
    b. outer race
    c. separator
    d. ball

2. What component of a typical ball bearing is indicated by the letter B in Figure 10-1?

    a. inner race
    b. outer race
    c. separator
    d. ball

3. What component of a typical ball bearing is indicated by the letter C in Figure 10-1?

    a. inner race
    b. outer race
    c. separator
    d. ball

4. Mechanic A applies lubricants to bearings only during assembly and mounting and during operation. Mechanic B also applies lubricants to bearings during storage and before removal. Who is right?

    a. Mechanic A
    b. Mechanic B
    c. Both A and B
    d. Neither A nor B

5. Too much lubricant can _____ .

    a. collect dirt
    b. cause overheating
    c. seep past seals and closures
    d. all of the above

6. Which of the following is an indication that premature bearing failure may occur?

    a. brinelling
    b. etching
    c. frettage
    d. all of the above

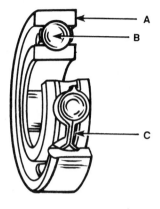

Figure 10-1

7. A vehicle's wheel bearings show signs of fatigue spalling. Mechanic A replaces the bearings. Mechanic B cleans them and reuses them. Who is right?

    a. Mechanic A
    b. Mechanic B
    c. Both A and B
    d. Neither A nor B

8. A vehicle's wheel bearings show signs of smearing of metal due to slippage. Mechanic A cleans the bearings and related parts. Mechanic B replaces the bearings, cleans related parts, and checks for proper fit and lubrication Who is right?

    a. Mechanic A
    b. Mechanic B
    c. Both A and B
    d. Neither A nor B

9. Usually, a seal's outside diameter should be _____ .

    a. 0.4 to 0.8 inch larger than the bore it goes into
    b. 1/4 to 1/8 inch larger than the bore it goes into
    c. 0.004 to 0.008 inch larger than the bore it goes into
    d. 0.1 to 0.4 inch larger than the bore it goes into

10. Mechanic A cleans bearings with steam. Mechanic B cleans bearings with water. Who is right?
    a. Mechanic A
    b. Mechanic B
    c. Both A and B
    d. Neither A nor B

JOB SHEET

SHOP ASSIGNMENT 35
REMOVE A BEARING
WITH AN ARBOR PRESS

NAME _____ STATION _____ DATE _____

Tools and Materials

Arbor press

Protective Clothing

Safety goggles or glasses with side shields

Procedure

1. Set up the arbor press to support the bearing while the press forces the shaft out of the bearing or to support the shaft while the bearing is being forced off the shaft.

Task completed _____

2. Does the bearing have a press-fit inner race?

Yes _____ No _____

   a. If no, go to step 3.

Not applicable _____ Task completed _____

   b. If yes, support the inner race on the press base plate with a bar or ring.

Task completed _____

   c. Using the press, apply force only to that race, forcing the shaft out of the bearing. (See Figure 10-2.)

Task completed _____

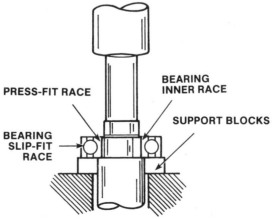

Figure 10-2

d. Go to the end to report any problems encountered.

Not applicable _____ Task completed _____

3. Does the bearing have two separable races, such as tapered roller bearings?

Yes _____ No _____

4. Loosen both races with the above procedure. (Never apply force to a slip-fit race or the cage.)

Not applicable _____ Task completed _____

PROBLEMS ENCOUNTERED: _____

_____

_____

INSTRUCTOR'S COMMENTS: _____

_____

_____

JOB SHEET

SHOP ASSIGNMENT 36
REMOVE FRONT-WHEEL
BEARINGS/SEALS

NAME _____ STATION _____ DATE _____

Tools and Materials

Hoist or jack stands
Wrenches
Wire, if needed
Pliers or screwdriver
Drift
New seal

Protective Clothing

Safety goggles or glasses with side shields

Procedure

1. Raise the front end of the vehicle on a hoist or safely support it on jack stands. (Do not support the vehicle on only a bumper jack.)

Task completed _____

2. Remove the hub cap or wheel cover (Figure 10-3).

Task completed _____

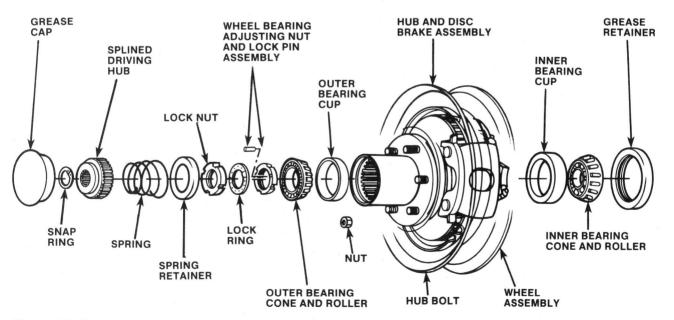

GREASE CAP

SPLINED DRIVING HUB

WHEEL BEARING ADJUSTING NUT AND LOCK PIN ASSEMBLY

HUB AND DISC BRAKE ASSEMBLY

GREASE RETAINER

INNER BEARING CUP

LOCK NUT

OUTER BEARING CUP

SNAP RING

SPRING

SPRING RETAINER

LOCK RING

NUT

OUTER BEARING CONE AND ROLLER

HUB BOLT

WHEEL ASSEMBLY

INNER BEARING CONE AND ROLLER

Figure 10-3

Chapter 10   Seal and Bearing Servicing

a. Use a wrench or jack handle to take off wheel lug nuts.

Task completed _____

b. Pull straight back to remove the wheel.

Task completed _____

c. On a vehicle with disc brakes, loosen and remove the brake caliper mounting bolts.

Not applicable _____ Task completed _____

d. Support the caliper while disconnected on the lower "A" frame or suspended by a wire loop.

Not applicable _____ Task completed _____

3. Use pliers or a screwdriver to remove the dust cover.

Task completed _____

a. Remove the cotter pin.

Task completed _____

4. Loosen the adjusting nut.

Task completed _____

a. Jerk the rotor or drum assembly to loosen the washer and outer wheel bearing.

Task completed _____

b. If step 4a is not done easily, the brakes might have to be backed off.

Not applicable _____ Task completed _____

c. Push the assembly back on the spindle.

Task completed _____

5. Remove the adjusting nut and washer.

Task completed _____

a. Remove the outer wheel bearing.

Task completed _____

6. Pull the drum or rotor assembly straight off the spindle, making sure the inner bearing or seal does not drag on the spindle threads.

Task completed _____

7. With seal side down, lay the rotor or drum on the floor.

Task completed _____

a. Place a drift or broom handle against the inner face of the bearing cone.

Task completed _____

b. Carefully tap out the old seal and inner bearing.

Task completed _____

8. Record the part number of the seal to aid in selecting the correct replacement.

Part no. _____ Task completed _____

a. Discard the old seal.

Task completed _____

9. Clean and inspect the old bearing thoroughly.

Task completed _____

10. Can the old bearing be reused?

Yes _____ No _____

a. If no, record the part number of the old bearing to ensure correct replacement.

Part no. _____ Task completed _____

PROBLEMS ENCOUNTERED: _____

_____

_____

INSTRUCTOR'S COMMENTS: _____

_____

_____

JOB SHEET

SHOP ASSIGNMENT 37
INSTALL FRONT-WHEEL
BEARINGS/SEALS

NAME _____ STATION _____ DATE _____

Tools and Materials

Grease
New seals
New bearings
Installation tools such as arbor press, drivers, press-fitting tools with an outside diameter approximately 0.010 inch smaller than bore size, soft striking mallet, and wood block and hammer if no other tools are available
Lint-free cloth
Clean paper
Bearing repacker, optional

Protective Clothing

Safety goggles or glasses with side shields

Procedure

1. Match part numbers to make sure the new seal is the correct one for the application.

   Task completed _____

2. By hand or with a bearing repacker, force grease through the cage and rollers or balls and on all surfaces of the bearing.

   Task completed _____

3. Place the inner side of the drum or rotor face up.

   Task completed _____

   a. Use drivers to drive the new cup into the hub.

   Task completed _____

4. Coat the hub cavity with the same wheel bearing grease to the depth of the bearing cup's smallest diameter.

   Task completed _____

   a. Apply a light coat of grease to the spindle.

   Task completed _____

5. Place the inner bearing on the hub.

   Task completed _____

   a. Lightly coat the lip of the new seal with the same grease.

   Task completed _____

6. Slide the seal onto the proper installation tool.

Task completed _____

　　a. Make sure the seal fits over the tool's adapter and the sealing lip points toward the bearing. (See Figure 10-4.)

Task completed _____

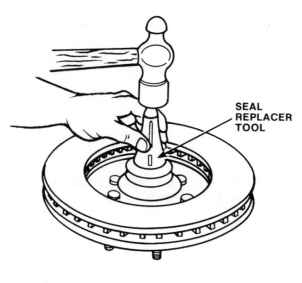

SEAL
REPLACER
TOOL

Figure 10-4

7. Position the seal so it starts squarely in the hub without cocking.

Task completed _____

　　a. Tap the tool until the seal bottoms out. (When the sound of the striking mallet changes, the seal will be fully seated in the hub.)

Task completed _____

8. If an installation tool is unavailable, use a wood block and hammer to drive in the seal. (Never hammer directly on the seal.)

Task completed _____

PROBLEMS ENCOUNTERED: _____

_____

_____

INSTRUCTOR'S COMMENTS: _____

_____

_____

CHAPTER 11

UNDERCAR FOUR-WHEEL DRIVE SERVICE

PRACTICE QUESTIONS

1. What type of front steering axle for a 4WD suspension is depicted in Figure 11-1?

   a. twin traction beam
   b. MacPherson suspension
   c. tubular beam
   d. freewheeling

2. What type of front steering axle for a 4WD suspension is depicted in Figure 11-2?

   a. twin traction beam
   b. MacPherson suspension
   c. tubular beam
   d. freewheeling

3. What type of front steering axle for a 4WD suspension is depicted in Figure 11-3?

   a. twin traction beam
   b. MacPherson suspension
   c. tubular beam
   d. freewheeling

4. What type of 4WD system has drive power to all the wheels all the time?

   a. Lo Lock
   b. freewheeling
   c. all-wheel
   d. LSD

5. After the wheel and tire combination on a 4WD vehicle is changed, the vehicle shows signs of steering linkage problems. Mechanic A attempts to solve the problem with add-on steering dampers. Mechanic B looks for defects in the tires. Who is right?

   a. Mechanic A
   b. Mechanic B
   c. Both A and B
   d. Neither A nor B

6. What component of a front steering axle is indicated by the letter A in Figure 11-4?

   a. sleeve
   b. wheel joint

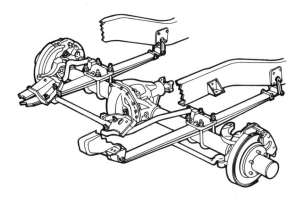

Figure 11-1

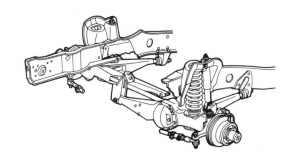

Figure 11-2

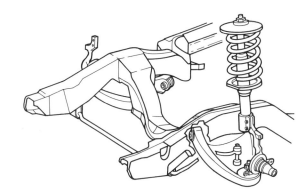

Figure 11-3

   c. ball joints
   d. knuckle

7. What component of a front steering axle is indicated by the letter B in Figure 11-4?

   a. sleeve
   b. wheel joint

c. ball joints

d. knuckle

8. What component of a front steering axle is indicated by the letter C in Figure 11-4?

    a. sleeve

    b. wheel joint

    c. ball joints

    d. knuckle

9. On-the-fly and automatic arrangements are slowly replacing what component to engage or disengage the axle shaft?

a. hub lock

b. radius arm

c. helper springs

d. staged shocks

10. Which of the following performance factors are affected when a lift kit is installed on a 4WD vehicle?

    a. the vehicle's center of gravity

    b. stopping distance and stability

    c. the geometry of the steering linkage

    d. all of the above

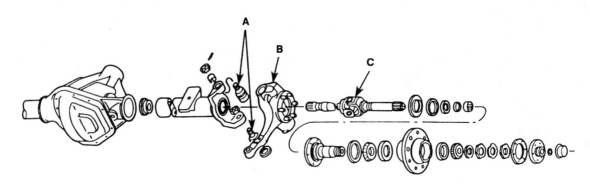

Figure 11-4

JOB SHEET

SHOP ASSIGNMENT 38
REPLACE A WHEEL LOCK HUB
ON A 4WD VEHICLE

NAME —————————— STATION —————————— DATE ————

Tools and Materials

Appropriate screwdrivers
Appropriate wrenches
Soft mallet
Grease
Torque wrench

Protective Clothing

Safety goggles or glasses with side shields

Procedure

1. Select a hub lock to fit the vehicle's hub design.

   Task completed ——

2. Referring to Figure 11-5, remove the hub cap, axle snap ring, drive
   flange, spring, and retainer from the internal hub.

   Not applicable —— Task completed ——

3. Remove the hub cap, axle snap ring, drive gear, and nut and washer
   from the external hub. (See Figure 11-5.)

   Not applicable —— Task completed ——

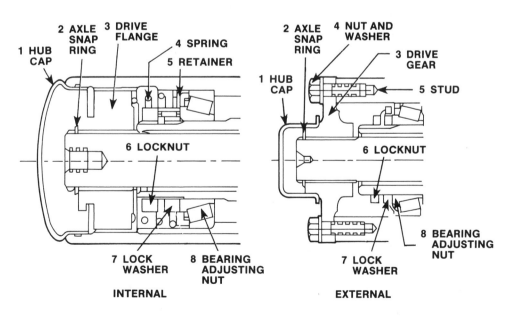

Figure 11-5

a. Remove the stud, if necessary.

Not applicable _____   Task completed _____

4. Carefully inspect the axle shaft splines to be sure they are free of nicks and burrs.

Task completed _____

5. Separate the base and handle units of the hub lock assembly.

Task completed _____

6. Apply a light coating of grease to the spline areas of the axle shaft.

Task completed _____

    a. Apply a light coating of grease to the hub lock base unit.

Task completed _____

    b. Apply grease to the O-ring.

Task completed _____

7. On external hubs, make sure the gasket surface is clean and smooth.

Not applicable _____   Task completed _____

    a. Use either the paper gasket furnished or RTV material, not thick, soft gasket material.

Not applicable _____   Task completed _____

8. Assemble the base unit to the wheel hub.

Task completed _____

9. On internal hubs, O-rings may require bumping with a soft mallet or the heel of the hand to get the base unit in place.

Not applicable _____   Task completed _____

    a. Lubricate the O-rings with a light coat of grease.

Task completed _____

10. Tighten the mounting bolts or stud nuts on external hubs to 25 foot-pounds in a cross pattern.

Not applicable _____   Task completed _____

11. Reinstall the snap ring on the axle.

Task completed _____

12. Push the hub lock handle into the base unit by hand. (This will be a tight fit due to the O-ring.)

Task completed _____

13. Align screw holes with the groove in the base unit.

                                                   Task completed _____

    a. Install four screws.

                                                   Task completed _____

    b. Do the screws enter freely?

                                                Yes _____   No _____

    c. If no, check for proper alignment.

                              Not applicable _____   Task completed _____

PROBLEMS ENCOUNTERED: _____

_____

_____

INSTRUCTOR'S COMMENTS: _____

_____

_____

JOB SHEET

SHOP ASSIGNMENT 39
INSTALL HELPER SPRINGS
ON A 4WD VEHICLE

NAME _____ STATION _____ DATE _____

Tools and Materials

Kit with all necessary hardware
Wrenches and other hand tools
Hoist
Floor stands

Protective Clothing

Safety goggles or glasses with side shields

Procedure

1. Raise the vehicle on a hoist.

   Task completed _____

   a. Support axle with floor stands.

   Task completed _____

2. Remove and discard the rear spring U-bolts.

   Task completed _____

   a. Remove the U-bolt guides and U-bolt retainer plate.

   Task completed _____

3. Place the helper spring on top of the supplied spacer block with the correct end of the spring to the rear of the truck and with the head of the center bolt in the hole of the spacer block.

   Task completed _____

4. Reuse the old U-bolt retainer plate and U-bolt guide.

   Task completed _____

   a. Install new U-bolts, washers, lock washers, and deep nuts.

   Task completed _____

5. Install the frame brackets, using the existing holes in the frame of the clamp wedges, if supplied, on the truck frame.

   Task completed _____

   a. Center the brackets over the end of the helper spring.

   Task completed _____

6. Tighten all the U-bolts to specifications.

Task completed _____

   a. Tighten the bracket nuts to specifications.

Task completed _____

7. Make sure the new spring is lined up properly to make contact with the brackets.

Task completed _____

8. Repeat procedure on other side of vehicle.

Task completed _____

9. Remove floor stands.

Task completed _____

   a. Lower the vehicle.

Task completed _____

PROBLEMS ENCOUNTERED: _____

_____

_____

INSTRUCTOR'S COMMENTS: _____

_____

_____

CHAPTER 12

TIRES AND WHEELS

PRACTICE QUESTIONS

1. What type of tire construction is depicted in the cutaway view in Figure 12-1?

   a. bias ply
   b. bias belted
   c. radial ply
   d. studded

2. Signs of improper tire inflation and uneven wear may indicate a need for _____ .

   a. balancing
   b. front suspension alignment
   c. rotation
   d. all of the above

3. Low tire pressure can result in _____ .

   a. tire bruising
   b. rapid wear on the outer edges
   c. carcass damage
   d. rapid wear at the center of the tire

4. A vehicle is equipped with radial tires that look like the tire depicted in Figure 12-2. Mechanic A says the bulge is normal for a radial tire. Mechanic B adds air to the tire. Who is right?

   a. Mechanic A
   b. Mechanic B
   c. Both A and B
   d. Neither A nor B

5. Which type of tire needs to be rotated the most often?

   a. bias ply and bias-belted
   b. radial
   c. both a and b
   d. neither a nor b

6. A vehicle's tires show signs of wear on one side. Mechanic A adjusts tire pressure. Mechanic B adjusts camber. Who is right?

   a. Mechanic A
   b. Mechanic B
   c. Both A and B
   d. Neither A nor B

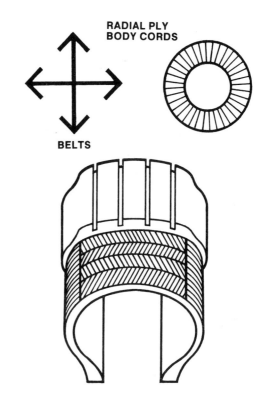

RADIAL PLY BODY CORDS

BELTS

Figure 12-1

Figure 12-2

7. Overinflation or lack of rotation can cause _____ .

   a. bald spots
   b. feathered edges
   c. rapid wear at the center
   d. cracked treads

8. What wheel dimension important to tire replacement is indicated by the letter A in Figure 12-3?

   a. offset
   b. diameter
   c. rim
   d. width

9. What wheel dimension important to tire replacement is indicated by the letter B in Figure 12–3?

 a. offset
 b. diameter
 c. rim
 d. width

10. Wheels that are statically unbalanced cause a bouncing action called _____

 a. wheel tramp
 b. runout
 c. offset
 d. none of the above

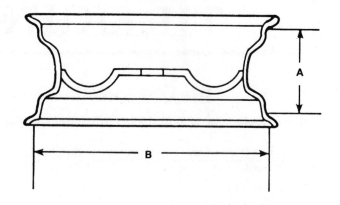

Figure 12-3

JOB SHEET

SHOP ASSIGNMENT 40
DISMOUNT A TIRE

NAME _____ STATION _____ DATE _____

Tools and Materials

Tire changer with manufacturer's instructions
Valve core removing tool

Protective Clothing

Safety goggles or glasses with side shields

Procedure

1. Remove the valve core to deflate the tire.

    Task completed _____

2. Place the narrow bead ledge of the rim up.

    Task completed _____

3. Following the tire changer manufacturer's instructions, force the tire bead inward from the wheel rim to unseat it.

    Task completed _____

4. Unseat the other bead.

    Task completed _____

    a. Hold one side of the bead in the drop center of the rim.

    Task completed _____

5. Remove the tire by placing the changer's lever or bar under the bead on the other side of the tire.

    Task completed _____

    a. Rotate the bar or lever around the rim to lift the tire bead to the outside of the wheel.

    Task completed _____

6. Repeat the process from the same side of the wheel for the opposite bead.

    Task completed _____

PROBLEMS ENCOUNTERED: _____

_____

_____

INSTRUCTOR'S COMMENTS: _____

_____

_____

JOB SHEET

SHOP ASSIGNMENT 41
MOUNT A TIRE

NAME _____ STATION _____ DATE _____

Tools and Materials

Tire changer with manufacturer's instructions
Rubber tire lubricant
Wire brush or coarse steel wool, if necessary
Nonabrasive cleaner, if necessary
Pulling tool
Wrenches and other hand tools
New valve stem and retaining nut, if necessary
Bead expander

Protective Clothing

Safety goggles or glasses with side shields

Procedure

1. Inspect the wheel.

   Task completed _____

   a. Is the wheel bent, dented, or heavily rusted? Does it have air leaks or elongated bolt holes or excessive lateral or radial runout?

   Yes _____ No _____

   b. If yes, replace the wheel.

   Not applicable _____ Task completed _____

2. Is the tire new?

   Yes _____ No _____

   a. If yes, determine if the valve stem is the snap-on type or if it has a retaining nut.

   Not applicable _____ Task completed _____

   b. If it is the snap-on type, install a new one from inside the wheel to the outside with a pulling tool.

   Not applicable _____ Task completed _____

   c. Make sure it is properly seated.

   Not applicable _____ Task completed _____

   d. If the stem has a retaining nut, remove it when pulling off the old stem.

   Not applicable _____ Task completed _____

e. Completely tighten the new nut.

Not applicable _____ Task completed _____

3. If the tire is not new, bend the valve stem to look for cracks.

Not applicable _____ Task completed _____

a. Is the rubber still supple?

Yes _____ No _____

b. If no, replace.

Not applicable _____ Task completed _____

c. Is the valve stem cracked?

Yes _____ No _____

d. If yes, replace.

Not applicable _____ Task completed _____

4. If the wheel is made of steel, clean the rim bead seats with a wire brush or coarse steel wool.

Not applicable _____ Task completed _____

a. If the wheel is aluminum, clean the rim bead seats with a nonabrasive cleaner.

Not applicable _____ Task completed _____

5. Lubricate the bead area with rubber tire lubricant.

Task completed _____

6. Use a tire changer and follow the manufacturer's instructions to guide the tire onto the wheel with the inner bead first.

Task completed _____

a. Hold one side of the inner bead into the drop center of the wheel to allow the other side of the inner bead to be rolled over the rim flange onto the wheel.

Task completed _____

b. Roll the outer bead over the wheel rim flange in the same way.

Task completed _____

7. Rotate the tire around the wheel until the valve stem is in line.

Task completed _____

8. Inflate the tire 10 psi or less, using a bead expander if necessary.

Task completed _____

a. Remove the expander.

Task completed _____

9. Are there signs of leakage?

Yes _____ No _____

a. If no, adjust air pressure to recommended pressure.

Not applicable _____ Task completed _____

b. If yes, repair the leak or dismount the tire.

Not applicable ⎯⎯ Task completed ⎯⎯

PROBLEMS ENCOUNTERED: ⎯⎯⎯⎯⎯⎯⎯⎯⎯⎯⎯⎯⎯⎯⎯⎯⎯⎯⎯

⎯⎯⎯⎯⎯⎯⎯⎯⎯⎯⎯⎯⎯⎯⎯⎯⎯⎯⎯⎯⎯⎯⎯⎯⎯⎯⎯⎯⎯⎯⎯⎯⎯⎯⎯

⎯⎯⎯⎯⎯⎯⎯⎯⎯⎯⎯⎯⎯⎯⎯⎯⎯⎯⎯⎯⎯⎯⎯⎯⎯⎯⎯⎯⎯⎯⎯⎯⎯⎯⎯

INSTRUCTOR'S COMMENTS: ⎯⎯⎯⎯⎯⎯⎯⎯⎯⎯⎯⎯⎯⎯⎯⎯⎯⎯

⎯⎯⎯⎯⎯⎯⎯⎯⎯⎯⎯⎯⎯⎯⎯⎯⎯⎯⎯⎯⎯⎯⎯⎯⎯⎯⎯⎯⎯⎯⎯⎯⎯⎯⎯

⎯⎯⎯⎯⎯⎯⎯⎯⎯⎯⎯⎯⎯⎯⎯⎯⎯⎯⎯⎯⎯⎯⎯⎯⎯⎯⎯⎯⎯⎯⎯⎯⎯⎯⎯

JOB SHEET

SHOP ASSIGNMENT 42
BALANCE REAR WHEELS ON A VEHICLE
WITH A LOCKING DIFFERENTIAL

NAME _____ STATION _____ DATE _____

Tools and Materials

Jack stands
Lug wrench

Protective Clothing

None required

Procedure

1. Raise the rear wheels by placing jack stands under the differential.

   Task completed _____

   a. Place the stands under the axle, but do not put any vehicle weight on the stands. (See Figure 12-4.)

   Task completed _____

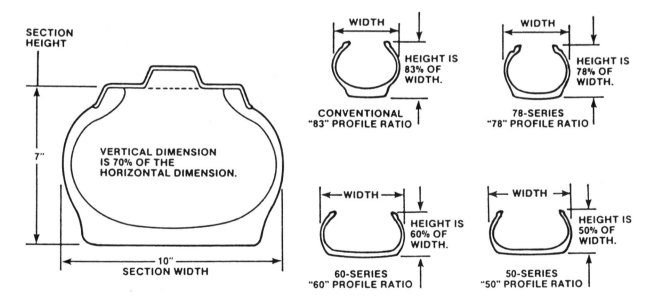

Figure 12-4

2. Remove one wheel.

3. Install the lug nuts.

    a. Tighten them to retain the brake drum.

4. Balance the remaining wheel using engine power to spin it.

5. Reinstall the wheel.

    a. Balance it.

PROBLEMS ENCOUNTERED: _____

_____

_____

INSTRUCTOR'S COMMENTS: _____

_____

_____

CHAPTER 13

UNDERCAR ALIGNMENT PROCEDURES

PRACTICE QUESTIONS

1. What component of a Rzeppa ball type fixed CV joint is indicated by the letter A in Figure 13-1?

   a. drive shaft
   b. ball cage
   c. boot
   d. outer race

2. What component of a Rzeppa ball type fixed CV joint is indicated by the letter B in Figure 13-1?

   a. drive shaft
   b. ball cage
   c. boot
   d. outer race

3. What component of a Rzeppa ball type fixed CV joint is indicated by the letter C in Figure 13-1?

   a. drive shaft
   b. ball cage
   c. boot
   d. outer race

4. What component of an outer tripod fixed CV joint is indicated by the letter A in Figure 13-2?

   a. thrust button
   b. boot
   c. tulip shaft
   d. outer race

5. What component of an outer tripod fixed CV joint is indicated by the letter B in Figure 13-2?

   a. thrust button
   b. boot
   c. tulip shaft
   d. outer race

6. What is the first component to be checked in a CV joint inspection?

   a. boots

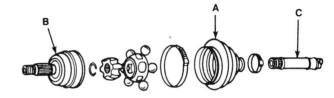

Figure 13-1

   b. bearing and bracket assembly
   c. drive shaft
   d. grease

7. Mechanic A uses special grease to lubricate CV joints. Mechanic B uses regular grease on CV joints. Who is right?

   a. Mechanic A
   b. Mechanic B
   c. Both A and B
   d. Neither A nor B

8. Separating the lower ball joint from the steering knuckle might require what tool?

   a. axle puller
   b. ball joint puller or fork
   c. impact wrench
   d. bearing remover

9. What tool can be used for staking a new nut?

   a. drift pin
   b. wrench
   c. dial indicator
   d. hardened chisel

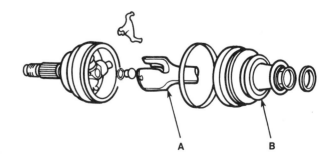

Figure 13-2

10. A vehicle with a half shaft damper is checked for excessive runout, and it is found that runout exceeds 1/4 inch. Mechanic A replaces the shaft. Mechanic B does not replace the shaft. Who is right?

    a. Mechanic A
    b. Mechanic B
    c. Both A and B
    d. Neither A nor B

JOB SHEET

SHOP ASSIGNMENT 43
REMOVE AN OUTER RZEPPA BALL TYPE CV JOINT
FROM HALF SHAFT

NAME _____ STATION _____ DATE _____

Tools and Materials

Soft jaw vise
Side cutters
Snap ring or duckbill pliers
Clean cloth
Soft hammer
Brass drift
Cleaning solvent
Constant velocity cleaner
Compressed air

Protective Clothing

Safety goggles or glasses with side shields

Procedure

1. Remove the half shaft from the vehicle.

   Task completed _____

   a. Place the half shaft assembly in a soft jaw vise.

   Task completed _____

2. Cut the clamps with side cutters.

   Task completed _____

   a. Remove the CV joint.

   Task completed _____

3. Wipe the grease away from the face of the joint.

   Task completed _____

   a. Is the joint secured by a snap or lock ring?

   Yes _____ No _____

   b. If yes, go to step 4.

   Not applicable _____ Task completed _____

   c. If no, remove the joint by tapping it off.

   Task completed _____

   d. Go to step 5.

   Not applicable _____ Task completed _____

4. If there is a snap ring, hold the tangs open with snap ring or duckbill pliers and strike the outer housing with a soft hammer to drive the joint off the shaft.

Not applicable _____ Task completed _____

   a. If necessary, use a brass drift against the face of the inner race to drive the joint off.

Not applicable _____ Task completed _____

5. Tilt the inner race to disassemble it.

Task completed _____

   a. Remove the CV balls in the necessary sequence.

Task completed _____

6. Tilt the inner race cross and cage 90 degrees to the outer housing to align the cage windows with the outer race.

Task completed _____

   a. Lift and remove the cage and inner race cross from the housing.

Task completed _____

7. Rotate the inner race upward and out of the cage.

Task completed _____

   a. Clean the components with solvent.

Task completed _____

   b. Dry with compressed air.

Task completed _____

   c. Rinse the CV joint with a constant velocity cleaner to completely remove residue.

Task completed _____

8. Inspect the joint for scoring or excessive wear on the bearings, cage windows, and inner and outer bearing races.

Task completed _____

   a. Examine for cracks, chips, or brinelling.

Task completed _____

PROBLEMS ENCOUNTERED: _____

_____

_____

INSTRUCTOR'S COMMENTS: _____

_____

_____

JOB SHEET

SHOP ASSIGNMENT 44
INSTALL AN OUTER RZEPPA
BALL TYPE CV JOINT

NAME _____ STATION _____ DATE _____

Tools and Materials

Soft jaw vise
Oil
Boot kit
Hand tools
New circlip, if necessary
Soft mallet

Protective Clothing

Safety goggles or glasses with side shields

Procedure

1. With the half shaft in the vise, inspect all components for reuse or replacement.

   Task completed _____

   a. Apply a light coat of oil on all parts to be used.

   Task completed _____

   b. Reassemble the wheel joint.

   Task completed _____

2. Align the inner race with the cage window.

   Task completed _____

   a. Rotate downward into the cage.

   Task completed _____

   b. Position the cage windows between the ball races and rotate downward.

   Task completed _____

3. Swing the race cross and cage 90 degrees in to the housing with the larger counterbore of the race cross facing outward.

   Task completed _____

4. Evenly distribute some of the grease that is provided in the boot kit into all of the ball bearing grooves.

   Task completed _____

5. Tilt the cage and race cross (with the bearing grooves and windows aligned) toward the ball grooves in the housing and insert the first ball.

Task completed _____

a. Insert the second ball the same way but in the opposite side.

Task completed _____

b. Install the rest of the balls in a similar fashion.

Task completed _____

6. Loosely place the new boot clamp on the half shaft.

Task completed _____

a. Carefully place the boot over the spline onto the shaft.

Task completed _____

b. Put electrical tape over the splines to prevent boot damage during installation.

Task completed _____

7. Install the new circlip from the kit or one provided.

Not applicable _____ Task completed _____

8. Position the wheel joint on the shaft splines.

Task completed _____

a. Use a soft mallet to sharply tap the hub nut joint onto the splined half shaft over the circlip.

Task completed _____

b. Pull lightly on the joint from the shaft to make sure that the circlip is seated correctly.

Task completed _____

9. Install the boot and a small clamp on the half shaft.

Task completed _____

a. Pack the boot with the remainder of the special grease.

Task completed _____

10. Make sure the large boot end is properly in place on the joint housing with no twists or crinkles appearing in the boot.

Task completed _____

a. Install and tighten (do not overtighten) both boot clamps.

Task completed _____

11. Flex the joints through their full range of motion to be sure they work smoothly.

Task completed _____

12. Install the half shaft on the vehicle.

Task completed _____

PROBLEMS ENCOUNTERED: _____

_____

_____

INSTRUCTOR'S COMMENTS: _____

_____

_____

JOB SHEET

SHOP ASSIGNMENT 45
REPLACE A DUST DEFLECTOR ON THE
INBOARD END OF A HALF SHAFT

NAME _____ STATION _____ DATE _____

Tools and Materials

Two special tools (shown in Figures 13-3 and 13-4)
Soft hammer
Hot tap water

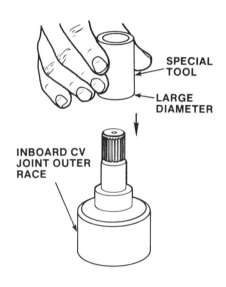

SPECIAL TOOL
LARGE DIAMETER
INBOARD CV JOINT OUTER RACE

Figure 13-3

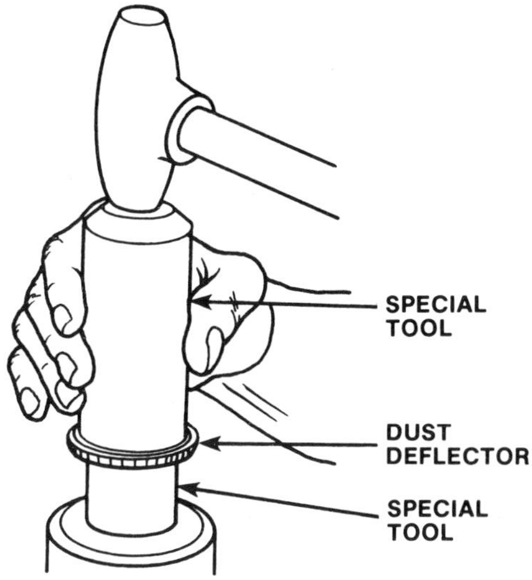

SPECIAL TOOL
DUST DEFLECTOR
SPECIAL TOOL

Figure 13-4

Protective Clothing

Safety goggles or glasses with side shields

Procedure

1. Remove the old dust deflector.

   Task completed _____

2. Soak the new deflector in hot tap water for a few minutes to aid installation and prevent splitting.

   Task completed _____

3. Place the special tool shown in Figure 13-3 over the inboard CV joint stub shaft with the large diameter of the tool toward the CV joint race.

   Task completed _____

4. Place the new deflector on the special tool with the dished edge facing upward and the ribbed face on the bottom.

Task completed _____

5. Use a tool similar to the one shown in Figure 13-4 to drive the deflector until it seats.

Task completed _____

PROBLEMS ENCOUNTERED: _____

_____

_____

INSTRUCTOR'S COMMENTS: _____

_____

_____

CHAPTER 14
UNDERCAR INSPECTION AND DIAGNOSIS

PRACTICE QUESTIONS

1. What is the first step in an undercar inspection?
   a. Make sure the vehicle curb height is correct.
   b. Check the friction ball joints and wheel bearings.
   c. Check the strut housing for damage or oil leakage.
   d. Inspect sway bars.

2. What measurement must be checked for an undercar analysis?
   a. caster
   b. camber
   c. toe-in
   d. all of the above

3. When inspecting the undercar of a vehicle with MacPherson suspension, Mechanic A raises the vehicle to check the upper strut mounts. Mechanic B checks the upper strut mounts before raising the vehicle. Who is right?
   a. Mechanic A
   b. Mechanic B
   c. Both A and B
   d. Neither A nor B

4. Which of the following vehicle components is not inspected during an undercar analysis?
   a. tires
   b. motor mounts
   c. carburetor
   d. CV joints

5. Rock the wheel from side to side and top to bottom to check _____ .
   a. shock-dampening capacity
   b. the condition of wheel bearings and friction ball joints
   c. coil springs
   d. none of the above

6. After conducting steering service to a vehicle, Mechanic A checks only toe. Mechanic B checks all alignment measurements. Who is right?
   a. Mechanic A
   b. Mechanic B
   c. Both A and B
   d. Neither A nor B

7. Which of the following is not a possible cause of vehicle vibration?
   a. loose or damaged engine mounts
   b. tire balance
   c. steering linkage problem
   d. universal or CV joint wear

8. Mechanic A checks yoke lash on a vehicle with rack-and-pinion steering at the wheel bearings. Mechanic B checks yoke lash at the struts. Who is right?
   a. Mechanic A
   b. Mechanic B
   c. Both A and B
   d. Neither A nor B

9. Which of the following noises are acceptable?
   a. camshaft belt whine
   b. noise from an automotive drive axle
   c. rear axle noise
   d. all of the above

10. Which of the following will contribute to an NVH problem when damaged or worn?
    a. U-joint
    b. rear chassis
    c. drive shaft
    d. all of the above

JOB SHEET

SHOP ASSIGNMENT 46
CONDUCT AN
UNDERCAR INSPECTION*

NAME _____ STATION _____ DATE _____

Tools and Materials

Car jack
Pry bar
Dial gauge, if necessary
Lift
Alignment tools and equipment

Protective Clothing

Safety goggles or glasses with side shields

Procedure

1. Make sure the vehicle curb height is correct.

Task completed _____

   a. Make a dry part check of the steering system.

Task completed _____

2. Rock the car at each corner to check the shock-dampening capacity.
(The car should not bounce more than once or twice after rocking each
corner.)

Task completed _____

3. Lift the car on a jack placed under the outer end of the control arm if
the car has springs seated on that arm.

Not applicable _____ Task completed _____

   a. If the car has springs seated on the upper control arm, place the
jack under the frame near the control arm.

Not applicable _____ Task completed _____

   b. Check the condition of the load-carrying joints.

Task completed _____

   c. Lift the wheel with a pry bar inserted between the bottom of the
tire and floor to check for excessive play in axial or vertical motion.

Task completed _____

4. Shake the tire from side to side to check radial or horizontal movement of the joint.

Task completed _____

*Any parts requiring servicing or replacing that are discovered during the inspection should be noted under
Problems Encountered.

a. If trouble is suspected in the joint, use an indicator tool such as a dial gauge for a more accurate examination.

Task completed _____

5. Is the joint equipped with a wear indicator?

Yes _____ No _____

a. If no, go to step 6.

Not applicable _____ Task completed _____

b. If yes, lower the vehicle and visually check the joint with the car resting on the wheels at the correct curb height.

Not applicable _____ Task completed _____

6. Does the car have a MacPherson or modified MacPherson suspension?

Yes _____ No _____

a. If no, go to step 7.

Not applicable _____ Task completed _____

b. If yes, lower the vehicle and raise the hood.

Task completed _____

c. Check the condition of the inner fender panel.

Task completed _____

d. Make sure that the upper strut mounts are securely fastened.

Task completed _____

7. Raise the car on a lift.

Task completed _____

a. Check the friction ball joints and wheel bearings.

Task completed _____

b. Rock the wheel from side to side and top to bottom to feel for play.

Task completed _____

c. Do the joints feel loose?

Yes _____ No _____

d. If no, go to step 7f.

Not applicable _____ Task completed _____

e. If yes, repeat step 7b, keeping an eye on the joint to observe movement.

Not applicable _____ Task completed _____

f. Check the rear wheel bearings on front-wheel drive cars.

Not applicable _____ Task completed _____

8. Examine the tires for unusual wear patterns and insufficient tread.

Task completed _____

9. Move from the tires to the control arms, checking for cracks or signs of damage.

Task completed _____

a. Check the mounting bushings at the inner end of the arms.

Task completed _____

10. Check coil springs for signs of damage or wear.

Task completed _____

a. Check to make sure the piston rod is not bent or coil ends have not broken off.

Task completed _____

11. On MacPherson suspensions, make sure the pinch bolt and the strut mounting area of the steering knuckle fittings are tight and serviceable.

Not applicable _____ Task completed _____

12. Check the strut housing for damage or oil leakage.

Not applicable _____ Task completed _____

a. Make sure the piston rod is not bent or cracked.

Not applicable _____ Task completed _____

b. Shake the strut assembly to make sure the upper strut mount and bearings are not worn, loose, or sagging.

Not applicable _____ Task completed _____

13. Visually inspect the tie-rod for bends or cracks.

Task completed _____

a. On both rack-and-pinion and parallelogram steering systems, check the outer tie-rod end by turning it.

Not applicable _____ Task completed _____

b. Does it turn stiffly, but smoothly?

Yes _____ No _____

c. If it turns too easily or binds in spots, it should be replaced.

Not applicable _____ Task completed _____

d. Is the inner tie-rod loose?

Yes _____ No _____

e. If yes, it should be replaced.

Not applicable _____ Task completed _____

14. On rack-and-pinion steering, check the inner tie-rod end by feeling the movement of the joint through the protective boot.

Not applicable _____ Task completed _____

a. Rock the wheel back and forth with the other hand while feeling the joint.

Not applicable _____ Task completed _____

15. Check the boots for tears, cracks, or missing clamps.

Task completed _____

a. Check for excessive dampness on or in the boot.

Task completed _____

16. Inspect sway bars, strut rods, and sway bar links for cracked or deteriorated bushings, bent bolts, or missing nuts and washers.

Task completed _____

a. Shake the strut rods to check for excessive movement.

Task completed _____

17. On parallelogram steering systems, inspect the idler arm, wearable pitman arms, and center link by pushing and pulling on the center link close to the idler arm.

Not applicable _____ Task completed _____

a. Check for excessive play in the parts.

Task completed _____

18. Check the rack mounting bushings and fittings on MacPherson suspensions to make sure they are serviceable.

Not applicable _____ Task completed _____

19. On power steering systems, check the hydraulic lines and vent tube to make sure that they are not bent, crimped, or otherwise damaged.

Not applicable _____ Task completed _____

20. On systems equipped with steering dampers, check the damper housing for signs of leaks, cylinder damage, and loose piston.

Not applicable _____ Task completed _____

21. On rack-and-pinion systems, check the yoke lash adjustment by grasping the pinion gear at the flexible coupling on the steering shaft and attempt to move it in and out.

Not applicable _____ Task completed _____

22. Check alignment.

Task completed _____

PROBLEMS ENCOUNTERED: _____

_____

_____

INSTRUCTOR'S COMMENTS: _____

_____

_____

JOB SHEET

NAME —————————— STATION —————————— DATE ————————

Complete the undercar analysis and alignment record in Figure 14-1, page 120.

UNDERCAR ANALYSIS AND ALIGNMENT RECORD					
Customer Name			**Make and Year**	**Mileage**	**Servicing Technician**
Phone No.			**Date of Service**		
Inspected Areas	**OK**	**Service Needed**	**Undercar Inspection Findings**	**Parts**	**Labor**
Alignment					
Tires					
Springs/Torsion Bars					
Shocks/Struts					
Ball Joints					
Control Arms and Bushings					
Strut Rod Bushings					
Stabilizer Links					
Pitman and Idler Arms					
Tie-Rod Linkage					
Rear Suspension					
Motor Mounts					
Universal/CV Joints					
Brake System					
Exhaust System					
Fluid Leaks					
R and P Mount Bush					
Inner Ends					
Outer Ends					
Bellows					
Rack and Pinion Unit			Subtotal		
Wheel Bearings			Total Estimate		

Alignment Record	**Measurements**	**Left Wheel**	**Right Wheel**
Caster	Initial		
	Factory Specification		
	Final		
Camber	Initial		
	Factory Specification		
	Final		
Toe-In	Factory Specification	Initial	Final

Figure 14-1

ANSWER KEYS

CHAPTER 1

1. b	3. d	5. c	7. c	9. a
2. a	4. b	6. b	8. d	10. b

CHAPTER 2

1. c	3. a	5. d	7. c	9. a
2. b	4. b	6. a	8. b	10. d

CHAPTER 3

1. d	3. b	5. d	7. a	9. a
2. b	4. c	6. d	8. b	10. c

CHAPTER 4

1. a	3. a	5. b	7. b	9. a
2. d	4. c	6. c	8. d	10. c

CHAPTER 5

1. b	3. a	5. b	7. a	9. b
2. d	4. b	6. d	8. c	10. d

CHAPTER 6

1. d	3. a	5. c	7. c	9. c
2. c	4. d	6. c	8. b	10. a

CHAPTER 7

1. d	3. b	5. d	7. d	9. c
2. a	4. c	6. b	8. a	10. a

CHAPTER 8

1. c	3. b	5. c	7. a	9. b
2. d	4. b	6. c	8. b	10. d

CHAPTER 9

1. a	3. c	5. c	7. d	9. b
2. b	4. c	6. d	8. a	10. a

CHAPTER 10

1. b	3. c	5. d	7. a	9. c
2. d	4. b	6. d	8. b	10. d

CHAPTER 11

1. c	3. b	5. a	7. d	9. a
2. a	4. c	6. c	8. b	10. d

CHAPTER 12

1. c	3. b	5. a	7. c	9. b
2. d	4. a	6. b	8. d	10. a

CHAPTER 13

1. c	3. a	5. b	7. a	9. d
2. d	4. c	6. a	8. b	10. a

CHAPTER 14

1. a	3. b	5. b	7. c	9. d
2. d	4. c	6. b	8. d	10. d